Amatterofdesign™

It's a matter of Packaging

Amatterofdesign™
It's a matter of Packaging

Published and distributed in Europe
and Latin America by

INDEX BOOK

Index Book S.L.
Consell De Cent 160 Local 3 08015 Barcelona Spain
Phone: +34 93 4545547 / +34 93 4548755 Fax: +34 93 4548438
Email: ib@indexbook.com URL: www.indexbook.com

viction:ary™

Published and distributed for
the rest of the world by viction:workshop

Rm2202 22/Floor Kingsfield Centre
18-20 Shell Street North Point Hong Kong
URL: www.victionary.com   Email: we@victionary.com

Edited and produced by viction:workshop

Book design by viction:design workshop
Art direction by Victor Cheung

ISBN 84-96309-01-0

Printed and bound in Hong Kong

Abbreviations

CD:	Creative Director
AD:	Art Director
D:	Designer
DA:	Design Assistant
PdM:	Production Manager
PjM:	Project Manager
CG:	Computer Graphic Designer
CW:	Copy-writer
I:	Illustrator
P:	Photographer
CC:	Copy and concept
CS:	Client supervisor
AE:	Account Executive

CONTENTS

Introduction 004

Interview #01 / K2design 006

Interview #02 / Sayuri Design 014

Interview #03 / Stylo Design 022

Section one: Visual 033

Section two: Touch 112

Section three: Idea 164

Contributors 220

Acknowledgements /
Future Editions 224

PG.04 / INTRODUCTION

When asked, 'What role does packaging play in your life?' many people might not be able to respond immediately or find it difficult to give a definite answer. However, if we think about the significance of packaging in our lives and how much it affects our decision-making process when we shop, it may surprise us. Many of us have probably bought one or more items simply for its attractive or fascinating packaging, regardless of the contents. At times when we are faced with several products that have similar contents and prices, we usually buy the one that has the most appealing packaging. We can test this theory by going into a supermarket and observing how people shop, especially parent and parents shopping with their children. On reflection, we might be able to understand why companies spend so much money packaging their products.

So what is the purpose or need of packaging? Is it to protect the product? Or is it simply a selling tool? When we trace back the origins of packaging, we can see that the basic reason for packaging was to protect, keep and carry. Food packaging was probably the most common. People made simple but effective packaging by using natural or raw materials like leaves, straw, wood and bamboo. This type of packaging may not have been advanced or flawless, but it certainly served its purpose. As our society and resources developed, the role of packaging design has also evolved. In recent years packaging design has flourished and undergone many changes, gone are the days when packaging was used purely as practical and informative purposes. Nowadays we live in an age where consumerism plays a crucial part in the society. Packaging has become an essential part of every company's marketing strategy.

Good packaging can enhance the image of a product and sometimes encourage impulse buying for consumers. So what can be categorized as good packaging design? As always, 'good' or 'bad' designs can be quite subjective and it is often a matter of taste and opinion. However with packaging design, most designers agree that successful designs should meet the needs of the consumers and the clients. But what about the designers' desires? Designers often aim to convey their creativity, individuality and style through their work, but due to constraints, it is not always easy to do so. If the clients are willing to let the designers be creative and inventive, the results can often be exciting and powerful. Of course, award-winning designs do not guarantee an increase in sales for all products, but the publicity may indirectly benefit the company or the product in the long run. The challenge for packaging designers is to produce creative work that can satisfy themselves, their clients and at the same time stimulate the consumers to purchase the products.

For many graphic designers packaging design often requires more consideration than other aspects of designs. Unlike editorial, corporate identity, web or advertising, most packaging is three-dimensional and technicality can be problematic. Designers do not just work on graphics alone; they also have to consider the construction, material used, production, durability, legibility and safety issues. Imagine designing a milk container for example. Since it will be sold on the shelves in supermarkets among maybe eight other similar products, it has to stand out and catch the consumers' attention. Apart from the graphics, the shape, the closure, the appropriate container material and durability for transport all have to be decided on. Remember the days when beverages only come in glass bottles? Now we can use materials like metal (can), paper, and plastic, but their environmental impact can be an issue, so the ability to recycle the packaging is an important factor too.

Yet not all packaging is used for commercial or practical reasons; in Japan both packaging and wrapping can be seen as an art form regardless of the purpose. The Japanese treat packaging and wrapping very seriously. There are books that teach people how to wrap, training courses for sales people or enthusiasts and thousands of wrapping paper and ribbons available for sale. Sales people in department stores wrap every purchase skillfully and carefully, sometimes advising you to return in thirty minutes so that they can finish wrapping. The Japanese believe that packaging or wrapping symbolizes respect of the buyer or giver. If the present is wrapped carelessly, the receiver may be offended by the lack of respect or gratitude. We have all experienced the anticipation and excitement when given a present that has been nicely wrapped. This feeling of joy and apprehension is sometimes more overwhelming than the seeing the unwrapped present.

This book, 'It's a matter of Packaging' is the fourth in the series of Amatterofdesign™. Like the three previous books: 'It's a matter of Editorial design', 'It's a matter of Identity' and 'It's a matter of Promotion', 'It's a matter of Packaging' invites readers on a journey around the globe that showcases the latest and most interesting packaging designs throughout the industry. The book is divided into three main sections: Visual, Touch and Idea, featuring designs from shopping bags to music, food and beverages, matches, books and toys. In 'Visual', there are emotive CD packaging by design teams, Zion Graphics and Non-Format. Zion Graphics not only use bright colours and bold typography, they have also abandoned printing on paper for plastic instead (p.60). Non-Format, on the other hand, use photographic imagery (a plastic robotic Adam from Garden of Eden) to provoke their viewers (p.72). Besides CD packaging, there is also Japanese design team Groovision, who were responsible for all the visuals within the 100% Chocolate Café in Japan (p.94). Their work includes designing paper cups, shopping bags, mugs, menus and chocolate wrapper.

In 'Touch', physical sensation becomes the subject matter. Groovision's Pelican tissue box is fun and refreshing, with not only graphics on the box but also on the tissue itself. (p.114). German/ Chinese designer Helen Fruehauf uses her cultural backgrounds to create a book that merges styles from the East and West. She uses traditional Japanese binding and different textiles that allow viewers to feel, which makes reading her book a truly unique experience (p.126). And London-based design team Form chose transparent plastic air bags to contain promotional items such as t-shirts and posters to send out to their contacts (p.131).

Boxes are used commonly in packaging, but they often get neglected after being used. However with intriguing concepts, plain boxes can be transformed into collectible items. In 'Idea', Japanese designer Yoshie Watanabe produced gift mirror boxes that create illusions for viewers. Each box has a leaf and a partially shown flower printed at the base and the use of mirrors enables viewers to see the entire flower as well as two leaves (p.166). Swiss design team Freitag use a humorous approach when designing their multi-purpose packaging/ boxes. The cardboard box not only shows a photo of the item inside, but it can be assembled into a fake TV set. They are also willing to send you a remote control if you return the response card (p.179)! Similar to boxes, label tags often get discarded too, but US-based design team Segura designed a set of tags for jeans that encourage customers to retain their tags for other uses. One of the tags can be converted into a carrier for I-Pod or PDA or mobile phone and another has a template with cut-outs for customers to play with (p.174).

Besides showcasing intriguing work from different designers around the globe, there is also a short biography of each designer whose work is featured in this book. And last but not least, there are interviews with three established and respected designers in the industry: Giannis Kouroudis from K2 design, Sayuri Shoji from Sayuri Studio and Tom Lancaster from Stylo Design. Although each designer is very different in style and cultural background, they each seek to maintain a balance between commercialism and creativity. Through their work, we can see that good designs can speak to us in all languages. Whether you are in the industry or not, this book will inspire, motivate and encourage you to observe what's around you. Most of all enable you to appreciate the art of packaging.

INTERVIEW WITH GIANNIS KOUROUDIS

Giannis Kouroudis graduated from University of Athens with a degree in Graphic Design and Photography. His designs and illustrations have been shown in three Biennale exhibitions, as well as being featured in many leading Greek and international journals.

Kouroudis is also the creative director of Korres, responsible for their corporate image. His work has won several awards in packaging design, corporate identity and advertising, including the EPICA Award for Illustrations & Graphics for "Athens 2004 Olympic Games Sports Pictograms".

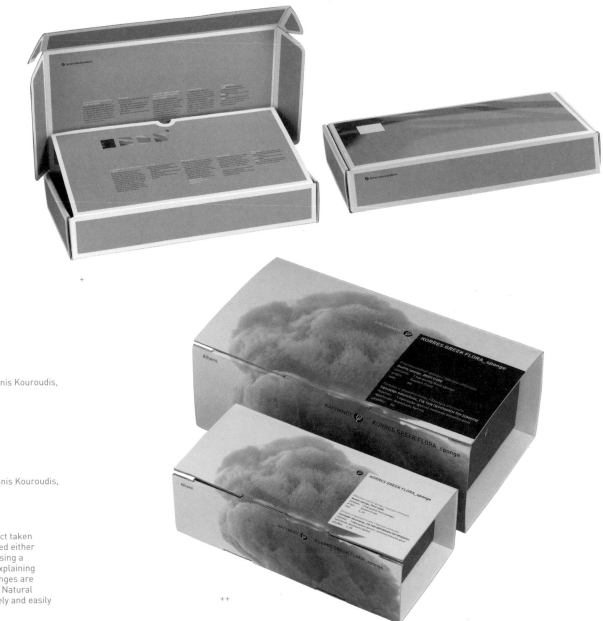

+ TITLE // Gift Box
TYPE OF WORK // Products' series
CLIENT // Korres Natural Products
YEAR PRODUCED // 2003
ART DIRECTION/DESIGN // AD: Yiannis Kouroudis,
D: Chrysafis Chrysafis
USE OF MATERIAL // Paper

++ TITLE // Natural Sponge Box
TYPE OF WORK // Products' series
CLIENT // Korres Natural Products
YEAR PRODUCED // 2003
ART DIRECTION/DESIGN // AD: Yiannis Kouroudis,
D: Chrysafis Chrysafis
USE OF MATERIAL // Paper

DESCRIPTION OF WORK //
Simple forms, pictures of the product taken on a coloured mirror and texts placed either horizontally or vertically on boxes using a simple font. Emphasis on the text explaining the origin and properties of the sponges are used in accordance with the Korres Natural Products aesthetics, which are widely and easily recognizable by the consumers.

+++ TITLE // Liquid Hand soaps
TYPE OF WORK // Products' series
CLIENT // Korres Natural Products
YEAR PRODUCED // 2003
ART DIRECTION/DESIGN // AD: Yiannis Kouroudis
USE OF MATERIAL // Recyclable plastic

DESCRIPTION OF WORK //
Simple forms, pictures of different herbs in their natural environment and texts placed either horizontally or vertically on boxes. Brochure containers employ a simple font alongside with an extensive use of white frames; all used in accordance with the Korres Natural Products aesthetics, which are widely and easily recognizable by consumers.

++++/+++++ TITLE // Korres Face Care
TYPE OF WORK // Products' series
CLIENT // Korres Natural Products
YEAR PRODUCED // 2003
ART DIRECTION/DESIGN //
AD: Yiannis Kouroudis, D: Chrysafis Chrysafis   +++
USE OF MATERIAL // Recyclable paper, aluminum, recyclable plastic

DESCRIPTION OF WORK //
Simple forms, extensive use of white background, pictures of different herbs placed on coloured mirrors and texts placed either horizontally or vertically on boxes. Brochure containers employ a simple font alongside with an extensive use of squares; all used in accordance with the Korres Natural Products aesthetics, which are widely and easily recognizable by the consumers.

+++++ TITLE // Toothpastes
TYPE OF WORK // Products' series
CLIENT // Korres Natural Products
YEAR PRODUCED // 2003
ART DIRECTION/DESIGN // AD: Yiannis Kouroudis,
D: Dimitra Diamanti
USE OF MATERIAL // Paper, plastic

DESCRIPTION OF WORK //
Simple forms, pictures of the products taken on a mirror and texts placed either horizontally or vertically on boxes. Containers employ a simple font, cubes and an explanatory text on the properties of the herbs used in the product.

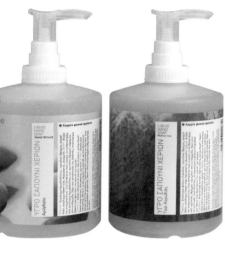

++++

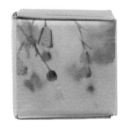

+++++

++++++

COULD YOU TELL US ABOUT K2DESIGN?

k2design is a firm that offers design solutions in communication problems. We believe that design apart from telling meaningful information can also bring art into our everyday life.

WHEN DID YOU SET UP YOUR DESIGN FIRM?

k2design was founded in 1988 by the name Kouroudis Design. The firm changed its name in 2002 because of the expansion of business and the introduction of new partners.

WHERE DO YOU FIND INSPIRATION? DOES YOUR COUNTRY INSPIRE YOU?

We have the privilege to live in a country with long history and distinguished civilization. This provides us with a stimulating everyday environment that forms the basis for new ideas.

WHAT DO YOU THINK ABOUT THE DESIGN CULTURE OF YOUR COUNTRY TODAY? DO YOU SEE ANY ADVANTAGES OR DISADVANTAGES BEING A DESIGNER IN GREECE?

Due to several historical circumstances, design in Greece is not as developed as in other European countries. However, the current design practice proves that new standards have been set up for further development. Athens 2004 Olympic Games was the reason for Greek design to become "the talk of the town".

WHAT IS YOUR DEFINITION OF PACKAGING?

We believe that packaging is the product's garment. Its purpose is to transfer the message of the product in the best possible way.

FROM YOUR POINT OF VIEW, WHAT PACKAGING DESIGN DO YOU THINK IS SUCCESSFUL?

Good packaging design should fulfill our clients' and their consumers' needs; it has to be unique and distinguishable, desirable and effective.

WHEN DID YOU START DOING THE CORPORATE IMAGE AND THE PACKAGING DESIGN FOR KORRES AND HOW?

We started our collaboration in 1996 when George Korres made his first steps. From the beginning, we worked together to set up the main principles (honesty, simplicity, aesthetic, scientific knowledge) of the company and then we developed a simple and flexible design structure.

COULD YOU TELL US ABOUT THE RELATIONSHIP BETWEEN YOU AND KORRES? ARE THEY YOUR MOTHER COMPANY OR THEY ARE JUST ONE OF YOUR CLIENTS?

Korres is not just one of our clients. He is one of the best ones. After nine years of close co-operation, George Korres has become a "trained" client with well-documented point of view. The trust and understanding that G.Korres has to our work contributes to the high quality design results.

DO YOU HAVE ANY GOOD OR BAD EXPERIENCE OF HANDLING THE PACKAGING DESIGN FOR KORRES?

The project difficulties can be creative and stimulating for the mind. Designing for Korres is always a creative and educational process, thus an enjoyable work.

WHEN YOU DID THE INTERIOR DESIGN OF THE KORRES' RETAIL STORES, WHAT WAS THE BIGGEST CHALLENGE YOU ENCOUNTERED?

For London's flagship store (2003), we collaborated with an English architectural office to create a unique shelfing unit and a design surrounding adjusted to brand values. In 2004, Barcelona's store followed the same principles but this time in collaboration with a Spanish architectural office. The biggest challenge we encountered was to adjust the established Korres brand to the international market. Fortunately, we did it successfully.

BESIDE THE KORRES' PROJECT, WHAT IS(ARE) THE FAVOURITE PACKAGING PROJECT(S) YOU HAVE EVER DONE? COULD YOU NAME A FEW OF THOSE? WHY?

We give our best all the time but I could mention the package for Mastihashop and the package for the oil called "Sitia". However, the project, that we are really proud of is the pictograms for the Athens 2004 Olympic Games.

WHAT CONTEMPORARY ARTISTS OR DESIGNERS HAVE RECENTLY GRABBED YOUR ATTENTION?

There are many artists whom I think highly of. Recently, Olafur Eliasson grabbed my attention with his weather project. Also, Erik Spiekermann of MetaDesign always does great work.

ARE YOU SELECTIVE ABOUT YOUR CLIENTS? WHAT KIND OF WORK OR TYPE OF CLIENT YOU ALWAYS EAGER TO HAVE?

We always want to have clients open-minded because the making process is a mutual "ride" in which both the client and the designer have to be ready for new and unexpected results.

ARE YOU SATISFIED WITH WHAT YOU ARE DOING NOW? WHAT WERE YOU EAGER TO DO IN YOUR CAREER PATH BUT HAVEN'T GOT THE CHANCE TO DO SO FAR?

We try not to be satisfied with ourselves in order to keep improving all the time by finding new ways of visualizing ideas.

WHAT IS YOUR ULTIMATE GOAL FOR K2DESIGN?

Our goal is to be always able to find the most appropriate way to communicate ideas but most of all we want to be happy with what we are doing.

www.korres.com

TITLE // Korres Barcelona store
TYPE OF WORK // Visual and graphics
CLIENT // Korres Natural Products
YEAR PRODUCED // 2003
ART DIRECTION/DESIGN // Yiannis Kouroudis

DESCRIPTION OF WORK //
Honesty, simplicity, aesthetic and scientific knowledge are the principles of
Korres products and we wanted to make them "visible" to the customers. This
store breathes confidence to the visitors and the soft grey design environment
prompts them to concentrate on the products. The photos that represent our
work are 7091 and 7092. The design of Korres Barcelona store is based on the
same concept as the London one (a big photograph on the wall opposite to the
one with the products' shelves), but this time in collaboration with a Spanish
architectural office.

TITLE // --
TYPE OF WORK // Packaging design
CLIENT // Mastiha shop
YEAR PRODUCED // 2003
ART DIRECTION/DESIGN // AD: Yiannis Kouroudis,
D: Dimitra Diamanti
USE OF MATERIAL // Design on traditional packages

DESCRIPTION OF WORK //
Mastiha shop has many products under its label. Our work
was to establish an unique and effective identity for them.
Mastiha is a traditional product and the base for many
goodies. We combined contemporary typeface with retro
postcards to show Mastiha's course over the years in the
best way.

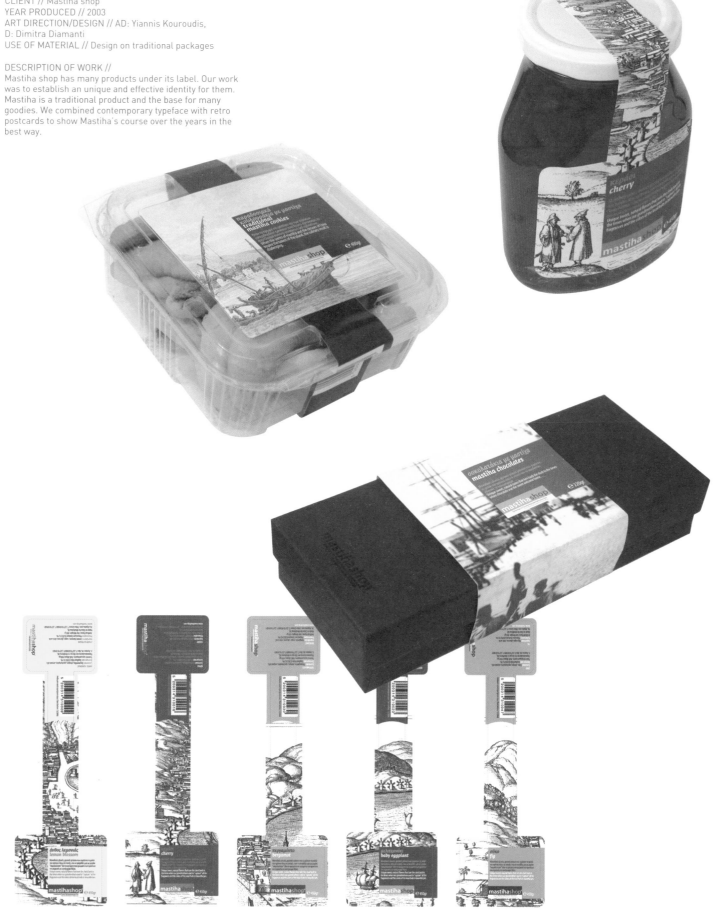

INTERVIEW WITH SAYURI SHOJI

Sayuri Studio, Inc. was founded in New York in 1998. The multi award-winning studio is known for its unique and highly original work including the revolutionary clear candle. They work for many international clients and they aim to develop concepts and designs that enhance their clients' images, as well as building value to their products.

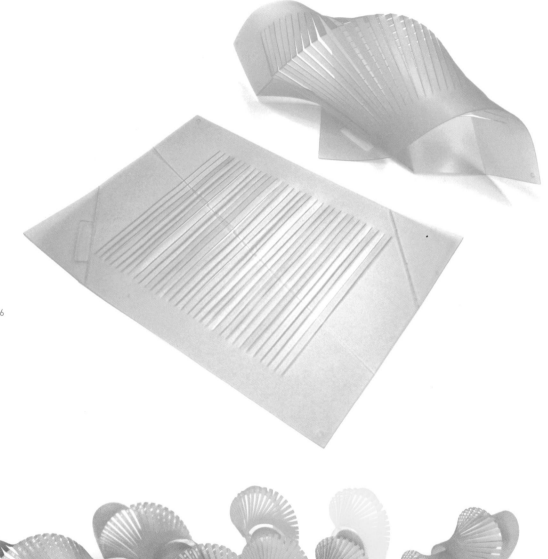

+ TITLE // Twistee 2004 Calendar
TYPE OF WORK // Product design
CLIENT // Sayuri Studio, Inc./Midori Japan
YEAR PRODUCED // 2003
ART DIRECTION/DESIGN // AD/D: Sayuri Shoji,
D: Yumico Ietsugu

DESCRIPTION OF WORK //
One, two, three steps and voila! A flat sheet becomes a three-Dimensional calendar! Make waves on your desk, or let it hang from your wall or ceiling! Enjoy the twelve colour variations every month. Copyrights: Sayuri Shoji PAT.P: 2002-383006

++ TITLE // Felissimo Shipping carton
TYPE OF WORK // Shipping carton
CLIENT // Felissimo Inc.
YEAR PRODUCED // 2002
ART DIRECTION/DESIGN // AD/D: Sayuri Shoji,
D: Yoshiko Shimozu

DESCRIPTION OF WORK //
The cartons feature a reversible interior to encourage customers to re-use them. Client can deliver many colourways to their customers every month (Felissimo is a mail catalog company) without changing the expensive gravure plates.

COULD YOU TELL US ABOUT SAYURI STUDIO?

We work on concept and product/ packaging design, fashion and beauty advertising for Japanese and global clients. Sayuri Shoji's early experiences include: in-house designer for Calvin Klein, art director for advertising agencies in New York City.

WHEN DID YOU SET UP YOUR DESIGN FIRM?

Established in New York City, moved studio to Tokyo since 2002

WHERE DO YOU GET INSPIRATION? DOES YOUR COUNTRY INSPIRE YOU?

I got most inspiration from my daily life. Not so much from design field... (People who I meet, Season, Nature, Street, Culture/ Art and so on) Some sense, yes, my country (now Japan) inspires me because I live here now. I strongly believe that design should be a rooted conversational business between you and who you communicate with.

JAPANESE PACKAGING DESIGN IS HIGHLY ACCLAIMED WORLDWIDE, WHAT DO YOU THINK ABOUT THE UPCOMING TENDENCY OF PACKAGING DESIGN IN JAPAN?

I personally think we (Japanese) are slight "design trend" victims. Too many trends, too many designs and products come and go and change in every minutes.

BEING A DESIGNER IN YOUR COUNTRY, DO YOU SEE ANY ADVANTAGES OR DIS-ADVANTAGES?

Advantage: Be able to have deep communication with clients & target consumers. Disadvantage: Often have to work on short cycled (turn around and perspective) projects.

I FOUND PACKAGING DESIGN AS ONE OF YOUR CORE BUSINESSES, HOW DOES THIS PATH EVOLVE?

5S project with Shiseido, and meeting my repre-sentative in NYC, Paul Meyers. He is my real sense partner since I became independent.

YOU HAVE DONE MANY REGIONAL AND OVERSEA PROJECTS, WHAT'S THE BIG-GEST DIFFERENCE BETWEEN WORKING WITH JAPANESE CLIENT AND OVERSEA CLIENT?

Of course there are some cultural differences between them. However, at basic concept I don't see a huge difference between domestic and over-sea clients. All clients love to see good design and want to be entertained by design!

YOU WORK WITH LOTS OF BIG CLI-ENTS LIKE ISSEY MIYAKE, HOW DID YOU START WITH THEM, AND ANY INTEREST-ING EXPERIENCE TO SHARE WITH US DURING THE DESIGN PROCESS?

I have many interesting and funny experiences with my clients... One thing I can say is that sometime I have opportunity to meet real great people or cli-ents whom you really can communicate with and they can bring me to next level.

WHAT IS(ARE) THE FAVOURITE PACKAG-ING PROJECT(S) YOU HAVE EVER DONE? COULD YOU NAME A FEW OF THOSE? WHY?

Pleats Please Issey Miyake shopping bag, Candle 0015/ 0017 and Felissimo Carton design, because they have new design value and have achieved strong presence. Recently I finished re-designing a green tea package for Kirin Beverage. This brand is one of their main businesses (Number two share in Japanese green tea market) and I found different perspectives on the design... This time I am con-vinced to sell over forty million cases! (One case= twenty-four)

COULD YOU SHARE WITH US WHEN YOU START DESIGNING THE PACKAGE FOR A PRODUCT, WHAT ARE THE CRUCIAL FACTORS DURING THE WHOLE PRO-CESS?

Two things. 1. Come up with concept. 2. Do best effort to achieve your goal within the budget, timing and technical reality.

WHAT IS YOUR DEFINITION OF PACKAG-ING?

To give clear "visible, attractive, distinctive" per-sonality to product.

COULD YOU NAME FIVE OF YOUR MOST FAVOURITE PACKAGING DESIGNS IN THE INDUSTRY?

Chanel No.5 perfume bottle, I-mac/ I-pod, Coca-cola, Marlboro.

WHAT CONTEMPORARY ARTISTS OR DESIGNERS DO YOU ADMIRE AND RESPECT?

Too many... (sorry, I can't name specific ones)

ARE YOU SATISFIED WITH WHAT YOU ARE DOING NOW? WHAT WERE YOU EAGER TO DO IN YOUR CAREER PATH BUT HAVEN'T GOT THE CHANCE TO DO SO FAR?

As a designer or person, I never get satisfied in a perfect way, and there are lots of things I haven't got the chance to do ... One sure thing I can say, I really love what I do, and what I can do is do my best on each opportunity I have today, that will open my future load.

WHAT IS YOUR ULTIMATE GOAL FOR SAYURI STUDIO?

Have fun and keep passionate!

TITLE // pleatsplease.com promotional CD
TYPE OF WORK // Promotional packaging
CLIENT // Issey Miyake, Inc.
YEAR PRODUCED // 2001
ART DIRECTION/DESIGN // Sayuri Shoji

DESCRIPTION OF WORK //
This promotional CD was made for the Pleats Please website launch. The rainbow bands graphic on the clear acetate echoes the product concept: "it's colorful, light and fun!" We also designed and art directed the website, and it won an award from CFDA (The Council of Fashion Designers of America), "The most-stylish.com".

www.pleatsplease.com

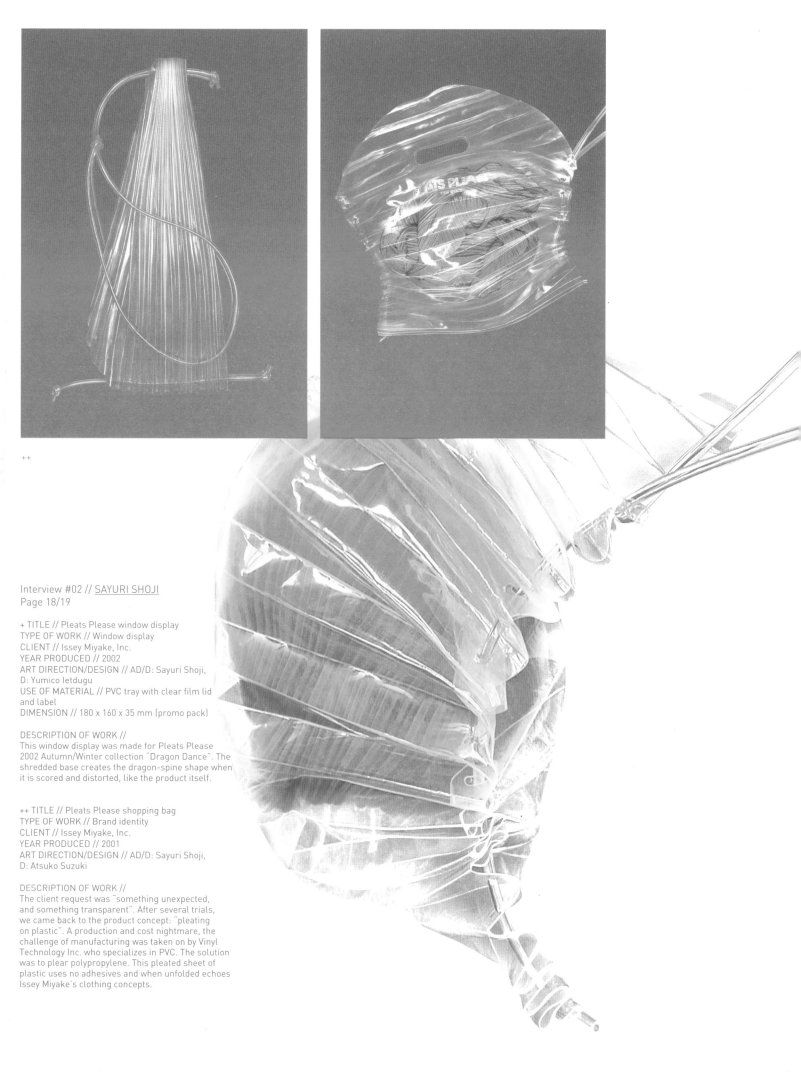

+ TITLE // Pleats Please window display
TYPE OF WORK // Window display
CLIENT // Issey Miyake, Inc.
YEAR PRODUCED // 2002
ART DIRECTION/DESIGN // AD/D: Sayuri Shoji,
D: Yumico Ietdugu
USE OF MATERIAL // PVC tray with clear film lid
and label
DIMENSION // 180 x 160 x 35 mm (promo pack)

DESCRIPTION OF WORK //
This window display was made for Pleats Please
2002 Autumn/Winter collection "Dragon Dance". The
shredded base creates the dragon-spine shape when
it is scored and distorted, like the product itself.

++ TITLE // Pleats Please shopping bag
TYPE OF WORK // Brand identity
CLIENT // Issey Miyake, Inc.
YEAR PRODUCED // 2001
ART DIRECTION/DESIGN // AD/D: Sayuri Shoji,
D: Atsuko Suzuki

DESCRIPTION OF WORK //
The client request was "something unexpected,
and something transparent". After several trials,
we came back to the product concept: "pleating
on plastic". A production and cost nightmare, the
challenge of manufacturing was taken on by Vinyl
Technology Inc. who specializes in PVC. The solution
was to plear polypropylene. This pleated sheet of
plastic uses no adhesives and when unfolded echoes
Issey Miyake's clothing concepts.

+ TITLE // H20 Fluid
TYPE OF WORK // Fragrance
CLIENT // H20
YEAR PRODUCED // 2001
ART DIRECTION/DESIGN // AD/D: Sayuri Shoji,
AD: Cindy Melk, D: Davd Seidler

DESCRIPTION OF WORK //
A toy and fragrance in one. Specially custom-tooled,
the outer layer carries the colored oil and the inner vial
holds the fragrance juice. It is really a fun fragrance to
play with and has a refreshing scent... It is flexible like
modern women today.

++ TITLE // CANDLE 0015/0017
TYPE OF WORK // Home/personal goods
CLIENT // Sephora USA, Sterling Group
YEAR PRODUCED // 2000
ART DIRECTION/DESIGN // Sayuri Shoji

DESCRIPTION OF WORK //
The first transparent solid wax candle ever on the
market. This aromatic candle is targeted at anyone who
appreciates a clean, modern style. "Clearer than ever"
is the product/packaging concept.

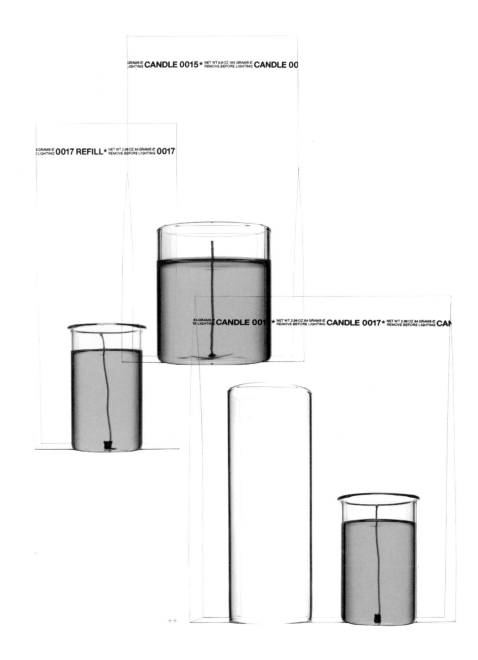

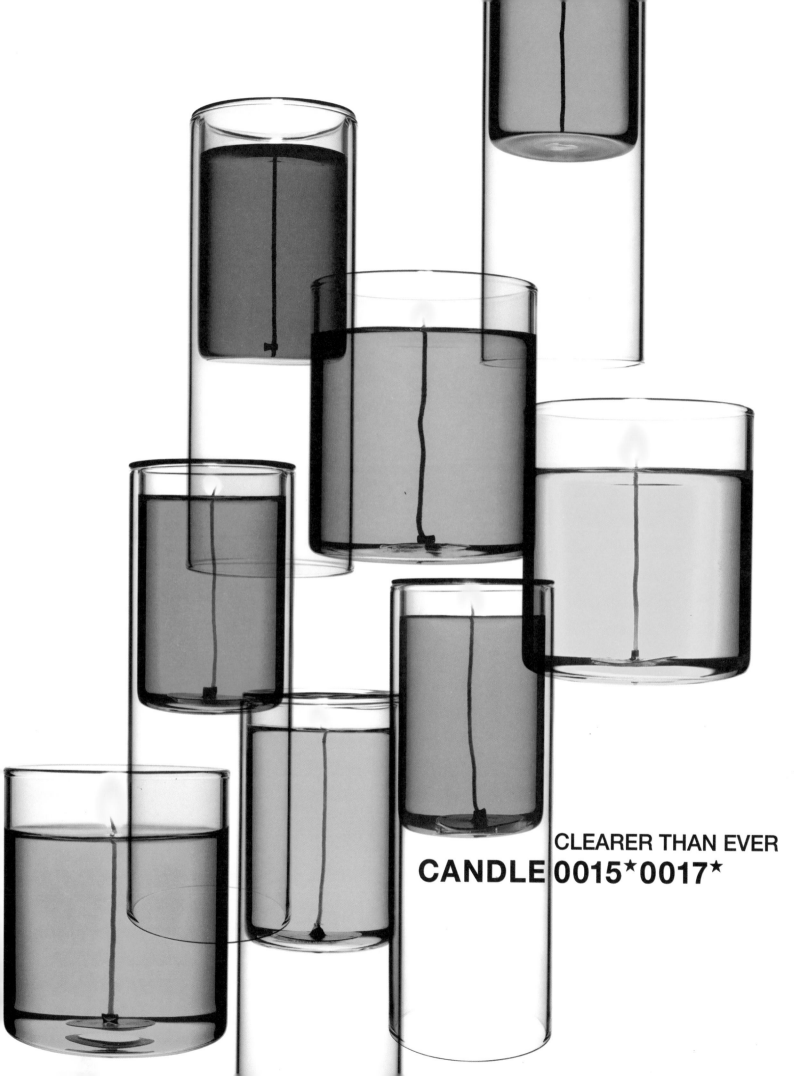

CLEARER THAN EVER
CANDLE 0015*0017*

INTERVIEW WITH TOM LANCASTER

Stylo Design is a creative design consultancy that works for public and private clients on a variety of projects including corporate identity, print, internet, e-commerce, moving image and sound design. Currently their team consists of five members, and each member has different skills and knowledge. This allows them to compliment each other and thus creating a wide range of work across different fields.

TITLE // British Beat
TYPE OF WORK // Compilation CD – Promotional and commercial release
CLIENT // Fly Records
YEAR PRODUCED // 2002
ART DIRECTION/DESIGN // Tom Lancaster
USE OF MATERIAL // PVC tray with clear film lid and label
DIMENSION // 180 x 160 x 35 mm (promo pack)

DESCRIPTION OF WORK //
Fly records is an independent record label who asked us to produce packaging for a compilation of best British releases. We produced packaging for promotional and commercial releases. We conceived the concept of 'British Beat', the design takes its cue from supermarket style cooked meat packaging design. The commercial release incorporates a slipcase printed with a 'meaty' image and the same supermarket packaging inspired graphics for the jewel case booklet and tray.

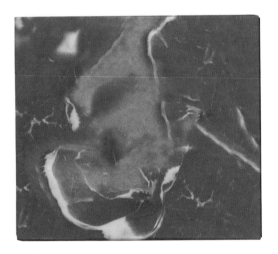

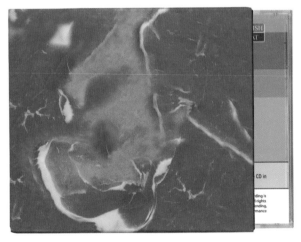

COULD YOU TELL US ABOUT STYLO?

Stylo was formed by myself (Tom Lancaster), my brother Ben Lancaster and Michael Williamson who was a colleague at Soho-based design consultancy Rise. Having been at Rise for five years, the time was right for all of us to set up our own venture. We wanted the challenge and creative control you get with running your own company. Our ethos is 'simple, quality, creative design'. We try to avoid 'over-design' in our work.

WHEN DID YOU SET UP YOUR DESIGN FIRM?

Stylo was officially established in 2001, although we had been working together as a collective on a variety of projects for the two years leading up to this as well.

WHY WOULD ALL YOUR PARTNERS HAVE AN IDEA OF GETTING TOGETHER AND SET UP A COMPANY THAT BELONGS TO YOU, AFTER EACH OF YOU HAVING WORKED LONG ON DIFFERENT FIELDS?

The timing was right for all of us and we felt that you don't pass up an opportunity to run your own company. We also had been building slowly to this point by collaboration, and so had laid the groundwork for launching the company.

We have different and complimentary skills, which allow us to take on a real broad cross section of work, be it web, print, identity, even sound design. Working on this variety of work keeps motivation high. We try and keep a good balance between the corporate and cultural arenas as it keeps work fresh. In our view, concentrating on just one of these sectors would be a mistake.

This has allowed us to develop a diverse portfolio (which continues to be so) of smaller creative through to larger more corporate projects.

WHAT IS YOUR DEFINITION OF PACKAGING?

Packaging is the consumer's first point of reference, it should be functional, protective, tactile and eye catching.

HOW ESSENTIAL IS A PACKAGING DESIGN IN OUR DAILY LIFE?

In very real terms packaging is not essential at all. As human beings I'm sure we can cope just fine without it. As a sales tool packaging can really help to make or break a product. A clever piece of packaging can become as important as the product itself.

WHERE DO YOU GET THE INSPIRATION FROM?

We get inspiration from living in London, arguably the best city for graphic design in the world. Also, there are so many quality design consultancies in London - Graphic Thought Facility, SEA, Intro, Form, Dixonbaxi, Farrow, Spin, Suburbia, Tom Hingston, Saville Associates, Poiint Blank Inc., Blue Source to name just a few. You can't help but take inspiration from others.

More generally we also take inspiration from all those things that as individuals make us tick. This is the usual type of thing; music, clothing, technology, art (in its many forms), humour etc.

IN YOUR OPINION, DO YOU THINK THAT THERE ARE DESIGN TRENDS IN DESIGN, NOT ONLY PACKAGING BUT ALSO ALL THE REST IN THE FIELD?

No question. Trends in design come and go as frequently as trends in any creative arena. The key is to be aware of them without facilitating them. We don't want to be the sort of consultancy that has a very distinct house style. We always evaluate a client's needs ahead of our own wishes to produce something creatively satisfying as if you are on your game, the end result should always provide this anyway. That said, clients would often approach us because they like the style of something we have designed in a particular discipline.

WOULD YOUR DESIGN IN CERTAIN WAY BE DRIVEN/AFFECTED BY DESIGN TREND?

We like to stay away from design trends. Invariably work produced with one eye on 'design trend' dates a hundred times faster than design produced with a more classic approach. Our design is a combination of client input and requirements, and an aesthetic evolved and refined over the years.

WHAT IS(ARE) THE FAVOURITE PACKAGING PROJECT(S) YOU HAVE EVER DONE? COULD YOU NAME A FEW OF THOSE? WHY?

We are particularly fond of the British Beat packaging we did for Fly Records. Humour in work is always a positive thing. This project utilized a pastiche of supermarket packaging for cold cooked meats.

Other favourites are the brochure housing we did for Foto (photographic agents) which replicated an old Second World War first aid tin. Only a small number of these tins were produced due to cost implications but they have been a constant source of curiosity among clients in the Foto office.

I FOUND YOU HAVE DONE LOTS OF CD PACKAGING. ARE YOU PARTICULARLY INTERESTED IN THIS AREA? WHAT CD PACKAGING DESIGN DO YOU THINK IS SUCCESSFUL FROM YOUR POINT OF VIEW?

It's not so much that we are particularly interested in this area but more that as the conventional replacement for vinyl, CD packaging has naturally become an area of focus for us. Design of CD packaging represents more of a challenge than vinyl due to the very obvious size restrictions. However budget allowing, great results can be achieved if you approach CD packaging from an unconventional perspective. This is the kind of CD packaging that is successful in our opinion. Some examples of this are Non-Formats packaging for Thurston Moore (Lo Recordings) - CD's packaged in vacuum cleaner bags filled with replica dust or Andy Muellers packaging for Pinebender (Ohio Gold Records) - a small pencil is included in the spine of the CD, the listener is encouraged to personalize their CD on graph paper cover of the Leaflet. Both of these examples can be found in CD-art by Charlotte Rivers (Rotovision).

ARE YOU SELECTIVE ABOUT YOUR CLIENTS? WHAT KIND OF WORK OR TYPE OF CLIENTS ARE YOU ALWAYS EAGER TO HAVE?

Not many design consultancies can afford to be selective about their clients. We only turn down work if the budget is ridiculously small. Even then we try and to suggest an alternative that can be produced within the client's financial constraints. We are always keen to keep our client base as diverse as we can. We like to work with clients who are simply nice people and therefore who we are motivated to develop partnerships with rather than just becoming a supplier. We also like to work with clients who have approached us due to previous work we have produced because you start a project half with a head start if you and the client are on the same wavelength.

HOW DO YOU CONVINCE THEM WITH YOUR IDEAS/SOLUTIONS?

We believe in managing expectation. We don't try and convince clients that a solution is the right one, unless they have specifically requested that we retain complete creative control. We always discuss ideas with a client and take on board what they want rather than approaching a job with the attitude that we know what's best for them. We work in partnership with our clients but provide them with consultancy based on our experience.

WHAT DO YOU THINK ABOUT THE PACKAGING DESIGN IN THE UK?

There's a lot of great packaging in the UK, from small scale to mass marketed products. An example of each: Mark Farrows pillbox for Spiritualized and Saturday's simple but effective packs for Boots own brand bathroom products, see http://www.saturday-london.com

ARE YOU SATISFIED WITH WHAT YOU ARE DOING SO FAR? WHAT ARE YOU EAGER TO ACHIEVE AT THIS MOMENT BUT DIDN'T GET THE CHANCE BEFORE?

Most designers are never 100% happy with their work on reflection. There will always be that little detail that needs tweaking but this is a critical element that drives us forward. It's a cliché but you really are only as good as your last project.

However, over the last six months or so we feel that we have reached a level where we are creatively satisfied and have a good flow of ongoing work with some great clients. We achieved one of our main goals running our own company and retaining a good level of creative control while also becoming financially stable. The ongoing challenge is to sustain and build on this.

WHAT IS YOUR ULTIMATE GOAL FOR STYLO?

Following on from the last question we want to maintain the way we work and the type of work we do and build on this. There cannot be one ultimate goal but rather a succession of goals on a weekly, monthly and yearly basis. We want to continue our controlled expansion of our client base, our team and our business.

+ TITLE // Double Knot
TYPE OF WORK // VHS slipcase series
CLIENT // Double Knot
YEAR PRODUCED // 2004
ART DIRECTION/DESIGN // Tom Lancaster
USE OF MATERIAL // Heavyweight recycled hard
board sleeves with laminate finish
DIMENSION // 105 x 190 x 30 mm

DESCRIPTION OF WORK //
Double Knot is a video production company
specializing in music videos. Stylo Design produced
a series of slipcase designs to sit alongside with the
identity using a simple solution of colour coding.

++ TITLE // Surefire Media Training DVD
TYPE OF WORK // DVD packaging
CLIENT // Surefire
YEAR PRODUCED // 2002
ART DIRECTION/DESIGN // Tom Lancaster
USE OF MATERIAL // Standard DVD case screen-
printed one colour and single colour disk
DIMENSION // 135 x 190 x 14.5 mm

DESCRIPTION OF WORK //
Innovative Media Training DVD – DVD packaging for
Surefire, a new media training company. Packaging
utilizes company branding and screen-printed single
colour on both packaging and DVD disk for a simple
but effective solution.

TITLE // Image Aid
TYPE OF WORK // Tin brochure housing
and brochure design
CLIENT // Foto
YEAR PRODUCED // 2003
ART DIRECTION/DESIGN // Tom Lancaster
USE OF MATERIAL // Screen-printed tin and
brochure printed on recycled stock with
laminate finish
DIMENSION // 145 x 215mm x 90mm (tin);
140 x 200mm (brochure)

DESCRIPTION OF WORK //
Foto represents a number of contemporary
photographers. We produced design solutions for
a range of printed marketing materials including
a portfolio brochure. The brochure concept takes
reference from imagery shot by photographers on
the agency's books. A one-off tin was produced
to house brochures, continuing the "Image Aid"
theme.

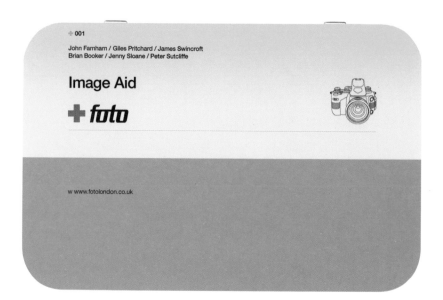

John Farnham / Giles Pritchard / James Swincroft
Brian Booker / Jenny Sloane / Peter Sutcliffe

Image Aid
Foto Photographic Ltd

✛
Foto Limited
29 Berringer Avenue
Brondesbury Park
London W9 8DN

t 020 8379
f 020 8379 422
w www.fotolondon.co.uk/imageaid
e contact@fotolondon.co.uk

✛ 003

John Farnham
Urban Environments
Signage London NW10

✛
w www.fotolondon.co.uk/imageaidFGP054
Image code No. FGP054

Giles Pritchard
Nature in focus

Giles Pritchard
Nature in Focus
Field View

✛
w www.fotolondon.co.uk/imageaidFGP002
Image code No. FGP002

✛ 007

John Farnham
Urban Environments
Autumn Night SW3

✛
w www.fotolondon.co.uk/imageaidFGP044
Image code No. FGP044

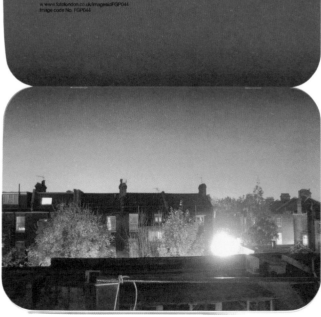

+ TITLE // Ladder
TYPE OF WORK // CD release
CLIENT // Lude Records
YEAR PRODUCED // 2003
ART DIRECTION/DESIGN // Tom Lancaster
USE OF MATERIAL // One colour screen-printed
polypropylene wallet and DVD case with housing for two CDs
DIMENSION // 170 x 240 mm (wallet), 135 x 190 mm (case)

DESCRIPTION OF WORK //
Lude Records commissioned us to design a promo release for
experimental group, Ladder. After listening to the music, we
took a simple illustrative approach with a mixed information
graphics as we felt it best represented the band. A screen-
printed plastic outer bag, two-colour CD pack and CDs were
produced.

++ TITLE // Surefire Media Training DVD
TYPE OF WORK // DVD packaging
CLIENT // Surefire
YEAR PRODUCED // 2002
ART DIRECTION/DESIGN // Tom Lancaster
USE OF MATERIAL // Standard DVD case screen-printed, one
colour and single colour disk
DIMENSION // 135 x 190 x 14.5mm

DESCRIPTION OF WORK //
Innovative Media Training DVD: DVD packaging for Surefire,
a new media training company. Packaging utilizes company
branding and screen-printed single colour on both packaging
and DVD disk for a simple but effective solution.

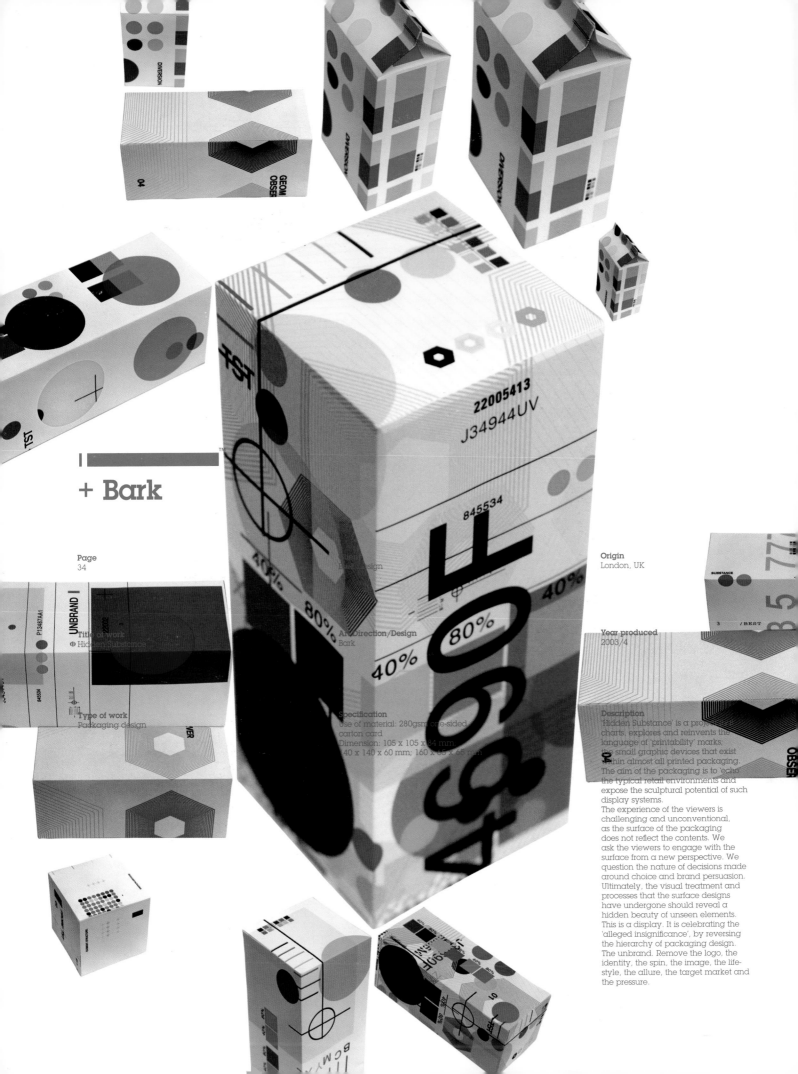

+ Bark

Page
34

Client
Bark Design

Art Direction/Design
Bark

Type of work
Packaging design

Specification
Use of material: 280gsm one-sided carton card
Dimension: 105 x 105 x 184 mm;
140 x 140 x 60 mm; 160 x 65 x 65 mm

Origin
London, UK

Year produced
2003/4

Description
'Hidden Substance' is a project that charts, explores and reinvents the language of 'printability' marks; the small graphic devices that exist within almost all printed packaging. The aim of the packaging is to 'echo' the typical retail environments and expose the sculptural potential of such display systems.

The experience of the viewers is challenging and unconventional, as the surface of the packaging does not reflect the contents. We ask the viewers to engage with the surface from a new perspective. We question the nature of decisions made around choice and brand persuasion. Ultimately, the visual treatment and processes that the surface designs have undergone should reveal a hidden beauty of unseen elements. This is a display. It is celebrating the 'alleged insignificance', by reversing the hierarchy of packaging design. The unbrand. Remove the logo, the identity, the spin, the image, the lifestyle, the allure, the target market and the pressure.

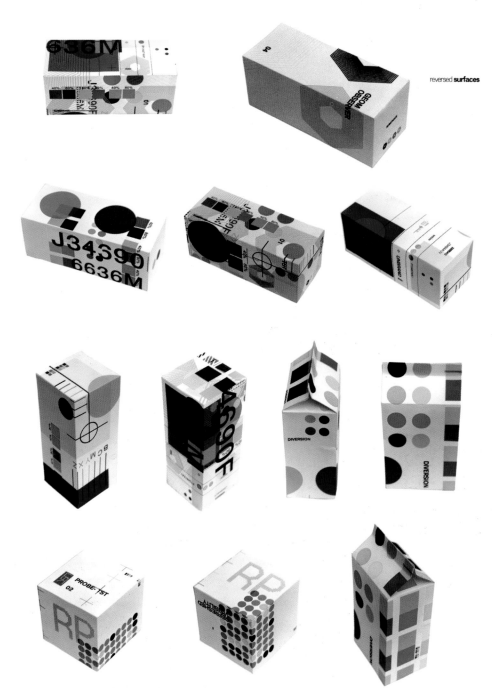

reversed **surfaces**

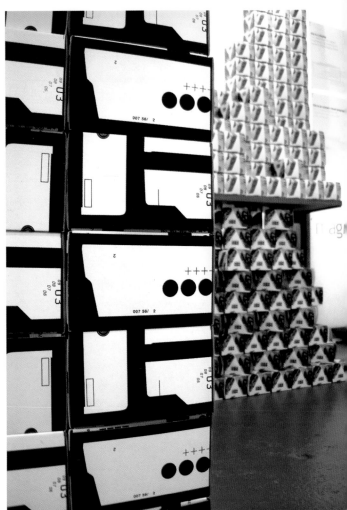

™

+ Yoshie Watanabe

Title of work
Dog bag

Type of work
Shopping bag

Client
Une Nana Cool Corp.

Art Direction/Design
Yoshie Watanabe

Origin
Tokyo, Japan

Year produced
2004

Specification
Use of material: Paper
Dimensions: 140 x 255 x 420 mm

Description

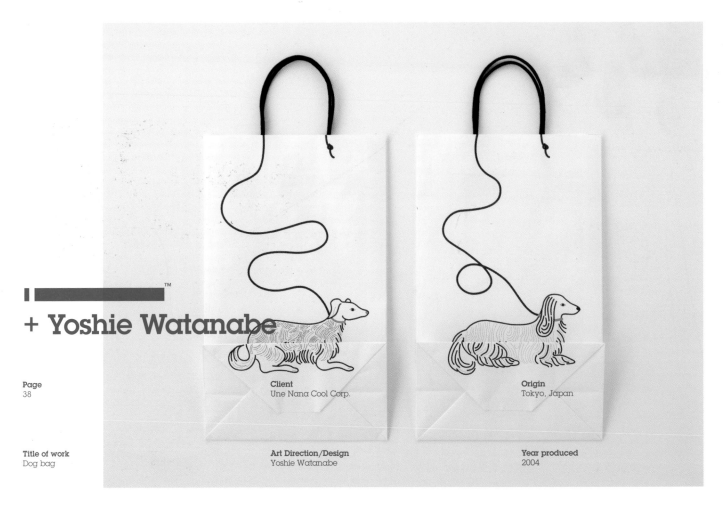

™

+ Bluemark

Page
39

Title of work
+ "mina perhonen" shopping bag
++ "National Standard" 2004-2005AW
shopping bag
+++ "National Standard" 2004SS
shopping bag

Type of work
Shopping bag

Client
+ MINA Co.,Ltd.
++/+++ National Standard Inc.

Art Direction/Design
AD/D: Atsuki Kikuchi

Specification
+ Use of material: Paper, craft band,
metal fitting offset & silk-screen print
Dimension: 420 x 380 x 130 mm;
260 x 240 x 90 mm
++ Use of material: Paper, craft band,
metal fitting offset print
Dimension: 300 x 340 x 120 mm;
260 x 240 x 100 mm
+++ Use of material: Paper, craft
band, metal fitting offset print
Dimension: 300 x 340 x 120 mm;
260 x 240 x100 mm

Origin
Tokyo, Japan

Year produced
+/+++ 2003
++ 2004

Description
+ These are shopping bags for "Mina
Perhonen", a Japanese fashion brand.
The butterfly pattern is the symbol of
the brand and it is printed on a white
silk-screen.
++ Every season, different patterns
are created for the fashion brand's
"National Standard" line. This season,
a logotype was printed that cannot be
seen from one side. A leopard pattern
was printed inside.
+++ For this season, an illustration
was designed to tell a story. Textured
paper was used to incorporate into the
overall feel of the image.

+ Form ™

Page
40

Title of work
+ Staverton carrier bag
++ The Moving Picture Company packaging
+++ Knoll tender folders
++++ Form brochure
+++++ Re:Creation House of Cards

Type of work
+ Bag
++ Two x carrier bags and one x video sleeve
+++ Ring-binder folder
++++ Brochure
+++++ Press release made up of a series of cards that slot into each other

Client
+ Staverton
++ The Moving Picture Company
+++ Knoll
++++ Form
+++++ Re:Creation Awards

Art Direction/Design
+ Paula Benson
++ Paul West
+++ AD: Paula Benson, D: Nick Hard
++++ AD: Paula Benson, Paul West, D: The Form studio
+++++ AD/D: Paul West, D: Paul West

Specification
+ Use of material: Litho printed card, matt laminated, die-cut handle
Dimension: 320 x 230 x 60 mm
++ Use of material: White foil block on white Paralux card
+++ Use of material: Screen-printed Polypropylene outer
Dimension: To house A4 sheets
++++ Use of material: Cover – Polypropylene, Inserts – Litho onto paper (Consort Royal era)
+++++ Use of material: Litho paper glued to card and die-cut

++

Origin
London, UK

Year produced
+ 1990
++/++++ 2003
+++/++++ 2004

++++

Description

+ This carrier bag was designed to be in keeping with the Staverton corporate identity – making strong use of the yellow. The bag was used for giveaways at furniture trade shows and launch events.

++ Carrier bag designed for this post-production house based in London's Soho district. In this area, lots of videos and CD's get distributed around town by runners and the bags are quite an important statement. However, rather than competing through being loud and brash, we decided to do the opposite and create a bag which stood out because of its subtlety and use of texture and typography. The handle shape echoes the MPC logo. The subtlety of the bag also encourages repeat use by the recipient.

+++ These ring-binder folders are screen-printed on the outside making strong use of the Knoll corporate red. The front is left predominantly blank so that the first inside page is visible through the plastic. The inside contains generic printed pages that can be customized in-house at Knoll, tendered to different types of clients.

++++ The brochure cover is a mini ring-binder made from polypropylene which has been screen-printed silver and die-cut with the Form dots logo to make the first page visible through the holes. The inserts feature different Form projects.

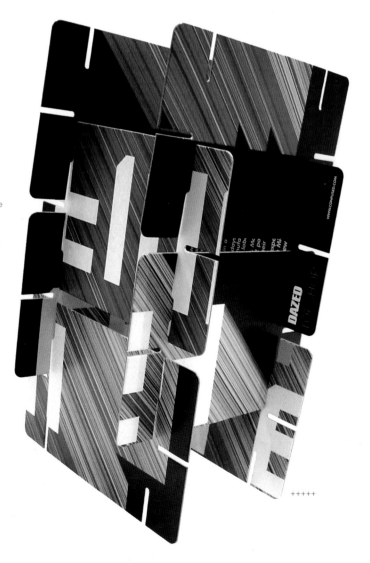

+++++

Page
42

Title of work
I'm feeling Wells

Type of work
Shopping bag

Client
Wells

Art Direction/Design
AD/D: Nuno Martins
D: Maria Koehler

Origin
Oporto, Portugal

Year produced
2004

Specification
Dimension: 46 x 47.5 x 14.5 cm

Description
Wells, a fashion accessorie company
increased its number of Po
stores in 2004. Their invest
used to create a new visua
for the brand and an aggr
communication strategy a
mainly at younger female

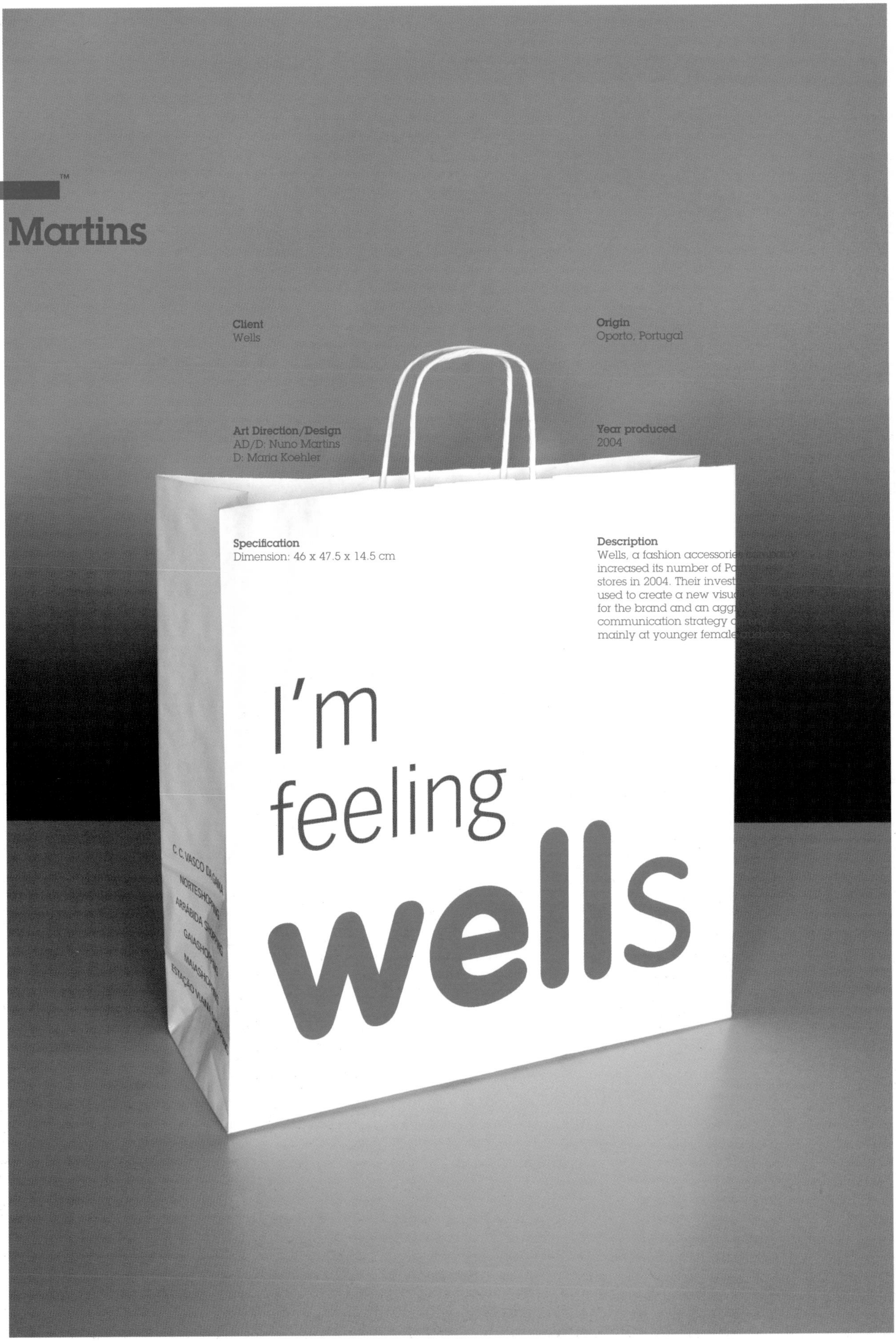

I'm
feeling
wells

C. C. VASCO DA GAMA
NORTESHOPPING
ARRÁBIDA SHOPPING
GAIASHOPPING
MAIASHOPPING
ESTAÇÃO VIANA

™

+ Teresa & David

Page
44

Client
Atlas

Origin
Stockholm, Sweden

Title of work
Kawaii

Art Direction/Design
Teresa & David

Year produced
2004

Type of work
Packaging, book and poster

Specification
Use of material: Paper, plastic
and bubble gum
Dimension: 240 x 160

Description
The book is about Japanese pop
culture, and the theme is Kawaii,
which means cute in Japanese.
Stickers and bubblegum were used,
alongside with a poster, plastic bag
and paper top.

+ butterfly-stroke inc.

Title of work
Enjoy 100%, Aiwa

Client
Sony Marketing

Art Direction/Design
CD/AD: Yukio Oshima
AD/D: Katsunori Aoki
CW: Kensho Yoshitani
I: Seijiro Kubo

Origin
Tokyo, Japan

Year produced
2002

Type of work
Package design

Specification
—

Description
Package design for a promotional campaign of an electronics company.

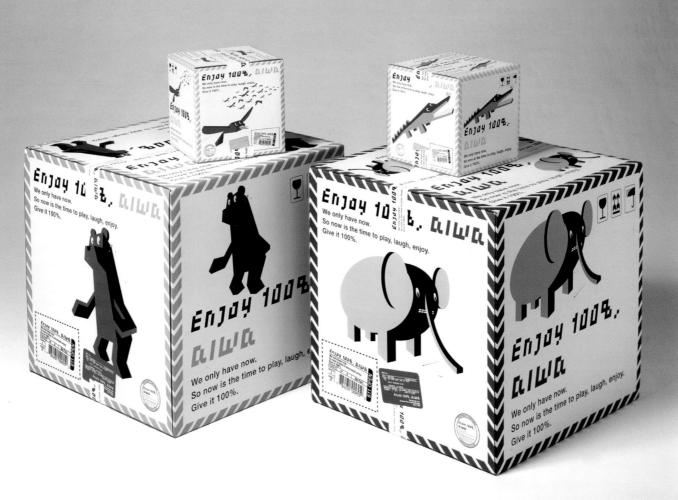

™

+ groovisions

Page
46

Title of work
+/++ Hacknet Logo
+++ Kazemachi Zukan/Takashi
Matsumoto
++++ Re./ram jam world
+++++ S.M.P.Ko2 STRIKES BACK!
(Second Mission Project ko2 Strikes
Back!)

Type of work
+/++ Shop item
+++ CD box design
++++ CD Jacket
+++++ Packaging design

Client
+/++ Memex Inc.
+++ Sony Music Entertainment
(Japan) Inc.
++++ WARNER MUSIC JAPAN INC.
+++++ Takashi Murakami/Kaikai
Kiki Co., Ltd.

Art Direction/Design
groovisions

Specification
+/++ Use of material: Paper, Plastic,
cardboard
Dimension: Paper bag (with red logo)
31 x 44 cm; Plastic bag 40 x 40 cm
+++ Use of material: Plastic, paper
Dimension: 10 x 13 x 14 cm
++++ Use of material: Paper,
bubble wrap
Dimension: Yellow 17.5 x 15.5 x 1 cm
+++++ Use of material: Cardboard
Dimension: 25 x 55 x 8 cm

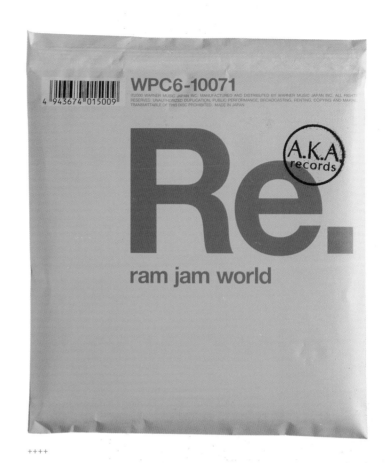

WPC6-10071

4 943674 015009

©2000 WARNER MUSIC JAPAN INC. MANUFACTURED AND DISTRIBUTED BY WARNER MUSIC JAPAN INC. ALL RIGHTS RESERVED. UNAUTHORIZED DUPLICATION, PUBLIC PERFORMANCE, BROADCASTING, RENTING, COPYING AND MAKING TRANSMITTABLE OF THIS DISC PROHIBITED. MADE IN JAPAN

Re.

A.K.A records

ram jam world

++++

+++

+++++

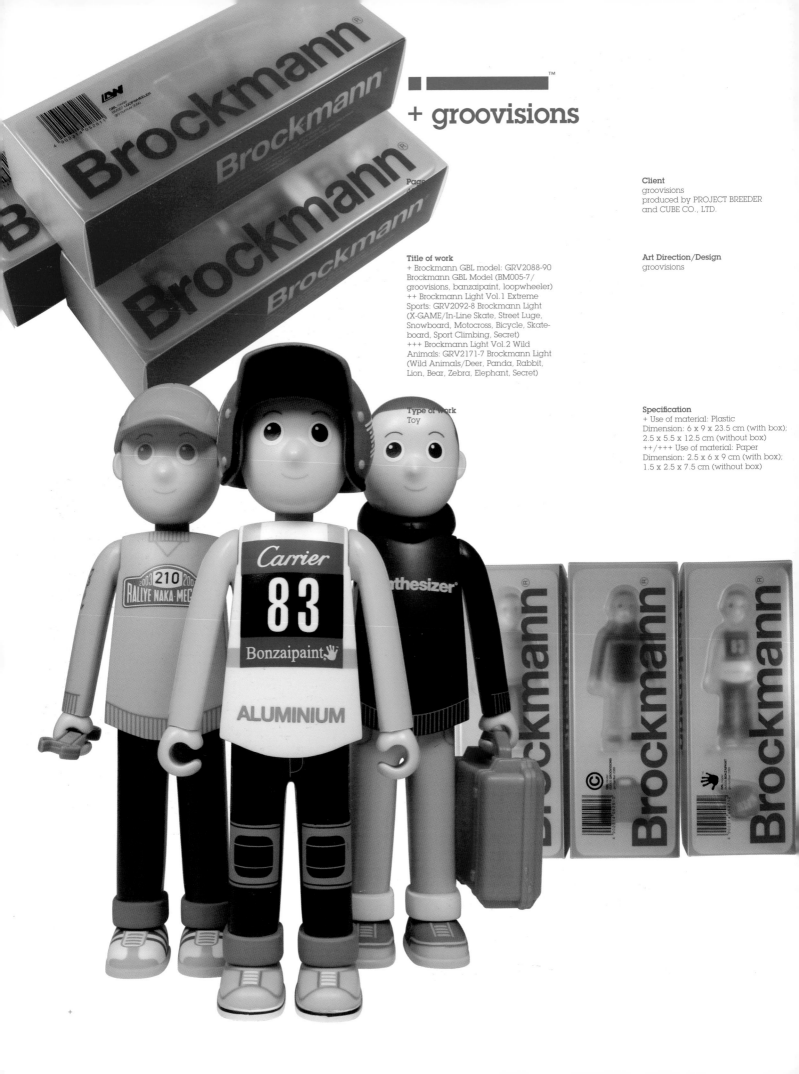

+ groovisions

Client
groovisions
produced by PROJECT BREEDER
and CUBE CO., LTD.

Title of work
+ Brockmann GBL model: GRV2088-90
Brockmann GBL Model (BM005-7/
groovisions, banzaipaint, loopwheeler)
++ Brockmann Light Vol.1 Extreme
Sports: GRV2092-8 Brockmann Light
(X-GAME/In-Line Skate, Street Luge,
Snowboard, Motocross, Bicycle, Skate-
board, Sport Climbing, Secret)
+++ Brockmann Light Vol.2 Wild
Animals: GRV2171-7 Brockmann Light
(Wild Animals/Deer, Panda, Rabbit,
Lion, Bear, Zebra, Elephant, Secret)

Type of work
Toy

Art Direction/Design
groovisions

Specification
+ Use of material: Plastic
Dimension: 6 x 9 x 23.5 cm (with box);
2.5 x 5.5 x 12.5 cm (without box)
++/+++ Use of material: Paper
Dimension: 2.5 x 6 x 9 cm (with box);
1.5 x 2.5 x 7.5 cm (without box)

Origin
Tokyo, Japan

Year produced
2004

Description
—

++

+++

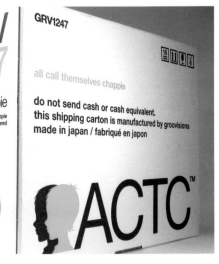

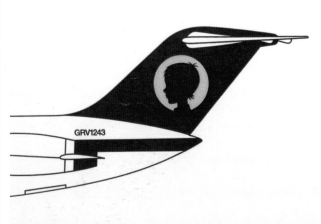

+ **groovisions**™

Client
+/++ Takashimaya
+++ groovisions

Origin
Tokyo, Japan

Title of work
+ GRV1247
++ GRV1252
+++ GRV1982 Box for T-shirts
(Orange, Lime Green, Silver)

Art Direction/Design
groovisions

Year produced
+/++ 1999
+++ 2003

Type of work
+/++ --
+++ Graphic for T-shirt and sweatshirt
box

Specification
+ Use of material: Cardboard
Dimension: 11 x 33.5 x 25.5 cm
++ Use of material: Cardboard
Dimension: 11 x 33 x.25.5 cm
+++ Use of material: Cardboard
Dimension: 37 x 26 x 2.5 cm

Description
--

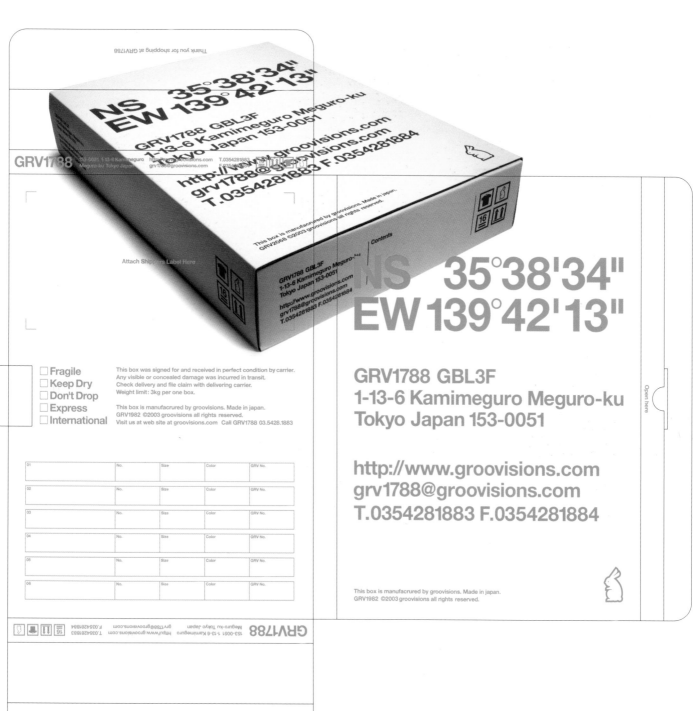

NS 35°38'34"
EW 139°42'13"

GRV1788 GBL3F
1-13-6 Kamimeguro Meguro-ku
Tokyo Japan 153-0051

http://www.groovisions.com
grv1788@groovisions.com
T.0354281883 F.0354281884

☐ Fragile
☐ Keep Dry
☐ Don't Drop
☐ Express
☐ International

This box was signed for and received in perfect condition by carrier.
Any visible or concealed damage was incurred in transit.
Check delivery and file claim with delivering carrier.
Weight limit: 3kg per one box.

01	No.	Size	Color	GRV No.
02	No.	Size	Color	GRV No.
03	No.	Size	Color	GRV No.
04	No.	Size	Color	GRV No.
05	No.	Size	Color	GRV No.
06	No.	Size	Color	GRV No.

+++

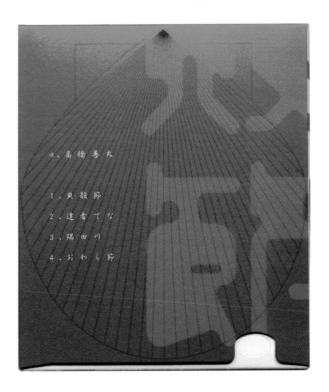

+ Kokokumaru

Type of work
CD package

Client
Paper Voice

Art Direction/Design
Yoshimaru Takahashi

Specification
Use of material: CD
Dimension: 140 x 124 x 7 m.

Page
53

Client
Farlove

Origin
Tokyo, Japan

Title of work
BOOK OF DAYS CD jacket

Art Direction/Design
AD/D: Atsuki Kikuchi

Year produced
2004

Type of work
CD cover

Specification
Use of material: Paper offset
& silkscreen print
Dimension: 125 x 140 mm

Description
This is a jacket of a compilation
album for the record label, Farlove.
The selection of music is arranged
under the themes: "the daily life"
and "everyday feeling", and warm
abstract forms are designed to
complement these themes. Matt and
risen ink was used in the silkscreen
printing.

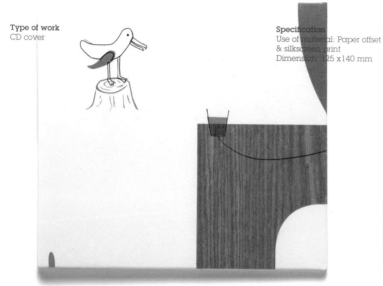

Origin
Osaka, Japan

Year produced
2004

Description
--

™
+ Ollystudio

Title of work
+ Disciple (grow old with me)
++ Pro-Keds European re-launch press
pack and invitations

Type of work
+ CD packaging and promotional
posters
++ Identity and packaging:
Invitations, posters, T-shirts and press
pack banners

Client
+ Cosmos Records
++ Pro-Keds

Art Direction/Design
AD: Oliver Walker
++ D: Simon Svärd

Specification
+ Use of material: Six-panel digipak,
foil blocked, jewel case
Dimension: 140 x 125
++ Use of material: Paper and board
Dimension: Various

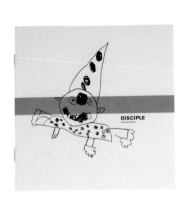

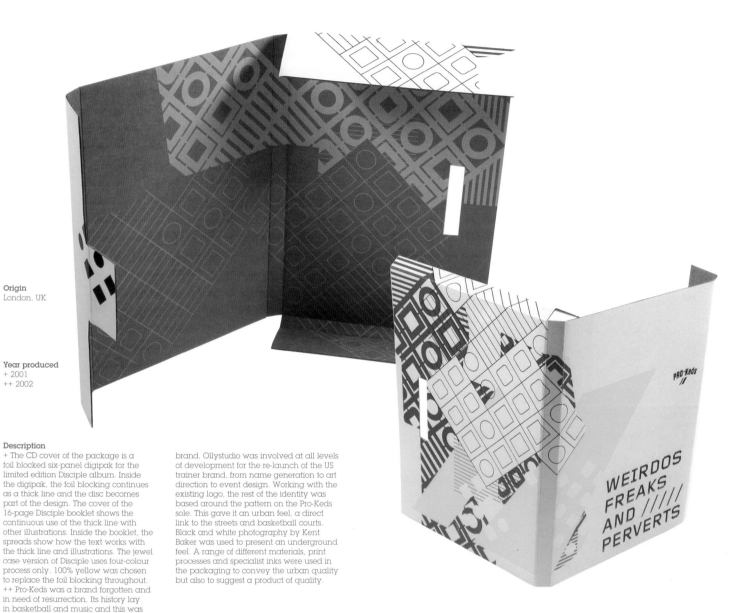

Origin
London, UK

Year produced
+ 2001
++ 2002

Description
+ The CD cover of the package is a foil blocked six-panel digipak for the limited edition Disciple album. Inside the digipak, the foil blocking continues as a thick line and the disc becomes part of the design. The cover of the 16-page Disciple booklet shows the continuous use of the thick line with other illustrations. Inside the booklet, the spreads show how the text works with the thick line and illustrations. The jewel case version of Disciple uses four-colour process only. 100% yellow was chosen to replace the foil blocking throughout.
++ Pro-Keds was a brand forgotten and in need of resurrection. Its history lay in basketball and music and this was the starting point for re-establishing the brand. Ollystudio was involved at all levels of development for the re-launch of the US trainer brand, from name generation to art direction to event design. Working with the existing logo, the rest of the identity was based around the pattern on the Pro-Keds sole. This gave it an urban feel, a direct link to the streets and basketball courts. Black and white photography by Kent Baker was used to present an underground feel. A range of different materials, print processes and specialist inks were used in the packaging to convey the urban quality but also to suggest a product of quality.

Page
56

Client
Plandx17

Origin
Compendium, Germany

Title of work
Merzbow's Tamago

Art Direction / Design
Bersin, Alf

Year produced
2004

Type of work
CD packaging

Specification
Use of material: Audio CD, ultra-white
cardboard (900G/m2 & matt), velcro
fastener, rubber
Dimension: 15 x 19.5 x 0.7 cm

Description
The CD packaging was created as a
limited collectors item for the album
"Tamago" by "Merzbow". The theme
was taken from the title "Tamago"
(egg) and was transformed into the
packaging concept. The design
reflects the minimalism and simplicity
of an egg.

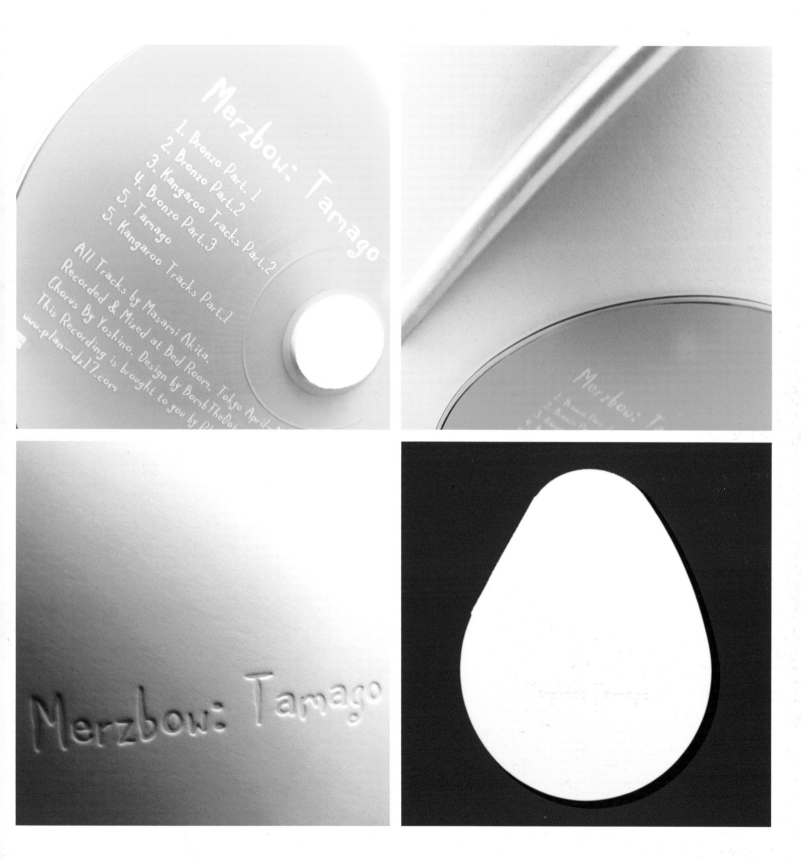

Merzbow: Tamago

1. Bronzo Part. 1
2. Bronzo Part.2
3. Kangaroo Tracks Part.2
4. Bronzo Part.3
5. Tamago
5. Kangaroo Tracks Part.1

All Tracks by Masami Akita.
Recorded & Mixed at Bed Room, Tokyo April
Chorus By Yoshino. Design by BombTheDo
This Recording is brought to you by Pl
www.plan-dx17.com

Merzbow: Tamago

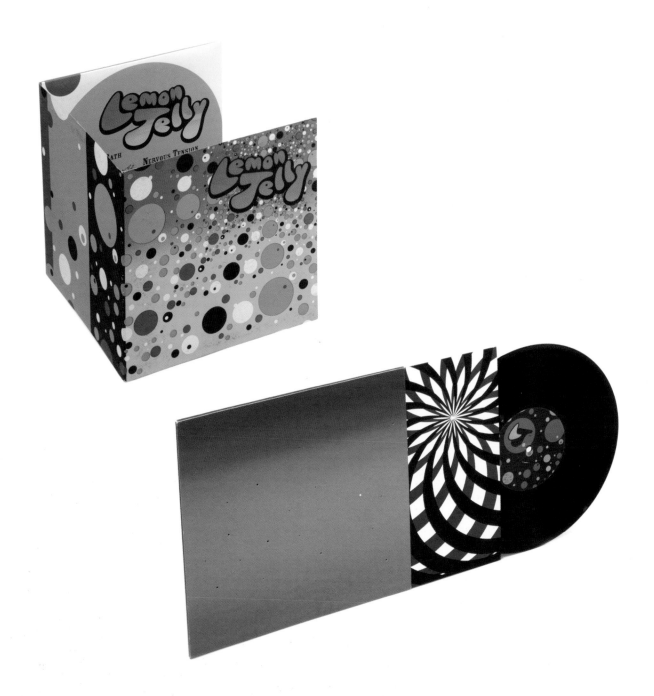
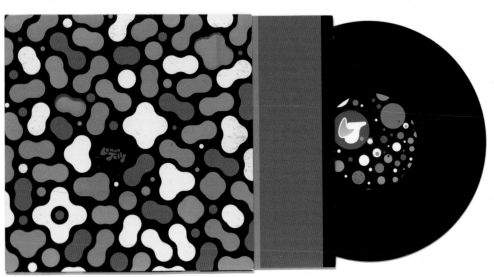

+ Airside™

Page
59

Client
Lemon Jelly

Origin
London, UK

Title of work
+ The Bath EP, The Midnight EP, The Yellow EP (from top left to bottom right)
++ Lost Horizons

Art Direction/Design
+ AD/D: Fred Deakin
++ AD: Fred Deakin
D: Airside

Year produced
+ Bath 1998, Midnight 2000, Yellow 1999
++ 2002

Type of work
Music packaging

Specification
+ Use of material: Reverse board + screen-printing
Dimension: 10"
++ Use of material: CMYK print on card
Dimension: 12"

Description
+ The Bath EP: Triple gatefold sleeve. The Midnight EP: The holes in the sleeve reveal a silver Lemon Jelly logo that glitters like stars as the inner sleeve is pulled out. Each cover is a unique hand modified gradient print. The Yellow EP: The cut-out shapes on the sleeve reveal an 'animated' logo on the inner sleeve which pops up in different places when the sleeve is pulled out.
++ The Lost Horizons artwork contains no type at all. The outer image shows beautiful rolling English green countryside alongside horrid grey cityscape. The same scene by night would be seen when the album is opened and suddenly the scene fades to blackness when it comes to life.

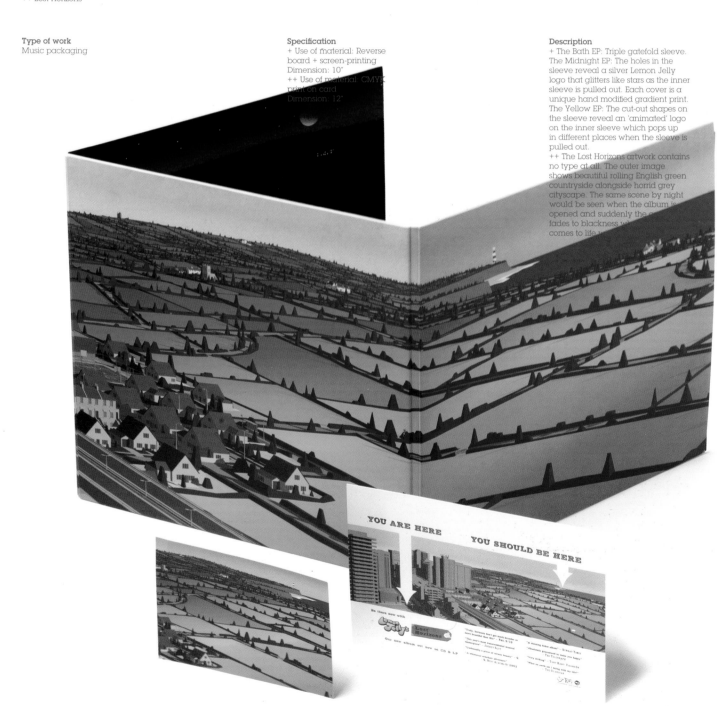

++

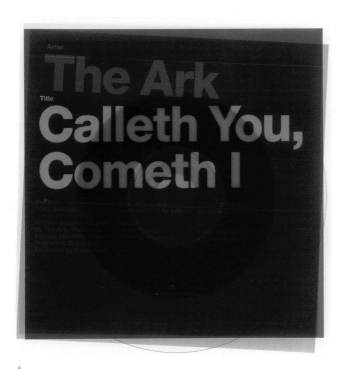

+ Zion Graphics

Page
60

Client
+ Virgin Records
++ BMG Music Publishing

Origin
Stockholm, Sweden

Title of work
+ The Ark: Calleth You Cometh I
++ Perfect Match

Art Direction/Design
Ricky Tillblad

Year produced
+ 2001
++ 2004

Type of work
+ Promo CD single
++ CD packaging

Specification
+ Use of material: Plastic
Dimension: 125 x 125

Description
+ CD promo for the band, The Ark.
The typography is split up on three
layers of plastic sheets in bright
colours and packaged in a clear
plastic sleeve. The CD is partially
see-through. The typography appears
differently depending on the order of
how the sheets are placed.

About This
Baxter

5050466-1122-2-4
S56 RECORDINGS

01	02	03	04	05	06
Got to Wake Up	Gonna Make it There	My Day	On My Own	Didn't Have a Choice	Can't
TIME: 3.52	TIME: 3.13	TIME: 3.51	TIME: 4.16	TIME: 4.13	TIME: 4.39
07	08	09	10	11	12
It's Coming	To You	I'm Here	Breathe in Breathe out	Get Done	Got to Wake Up (Fleshquartet De-Mix)
TIME: 4.45	TIME: 5.45	TIME: 2.44	TIME: 4.17	TIME: 4.05	TIME: 4.13

Written, performed and produced by Baxter. Mixed by Carl-Michael Herlöfsson for Primal Prod. at Primal Studio. Close-up photos by Peter Gehrke. Cover by Zion Graphics. Mastered by Björn Engelmann at Cutting Room. Web prod. by 24HR.

Experience it at: www.baxtered.com

Additional musicians: Goran Kajfes; Trumpet on track 03, Christian Hörgren; Cello on track 07, Patrick Andersson; Pedal steel on track 07. (Herlöfsson, Tillblad published by Primal Publishing/Air Chrysalis Scand., Ramsby, Copyright Control).

℗ & © 2002 S56 Recordings AB

www.s56.com

5050466-1122-2-4

Got to
Wake Up
Baxter

01	02
Got to Wake Up	Got to Wake Up (Fleshquartet De-Mix)
TIME: 3.52	TIME: 3.13

Taken from the forthcoming album "About this"

Written, performed and produced by Baxter. Mixed by Carl-Michael Herlöfsson for Primal Prod. at Primal Studio. Close-up photo by Peter Gehrke. Cover by Zion Graphics.

Mastered by Björn Engelmann at Cutting Room. (Herlöfsson, Tillblad published by Primal Publishing / Air Chrysalis Scand., Ramsby, Copyright Control).

℗ & © 2002 S56 Recordings AB. www.s56.com

ECLECTIC BOB** Chocolate Garden

→ PRIMCD 006

Page
63

Client
+/++ s56 Recordings
+++ Primal Music

Origin
Stockholm, Sweden

+ Zion Graphics

Title of work
+ Baxter: About This
++ Baxter: Got To Wake Up
+++ Eclectic Bob: Chocolate Garden

Art Direction/Design
Ricky Tillblad

Year produced
+/++ 2002
+++ 2001

Type of work
CD packaging

Specification
+ Use of material:
UV-varnished paper in jewel case
Dimension: 120 x 120
++ Use of material:
UV-varnished cardboard
Dimension: 125 x 125
+++ Use of material: Paper in
jewel case
Dimension: 150 x 125

Description
–

+ Zion Graphics

Page
64

Client
s56 Recordings

Origin
Stockholm, Sweden

Title of work
+ Thomas Rusiak: In The Sun
++ Thomas Rusiak: Unicorn

Art Direction/Design
Ricky Tillblad

Year produced
2003

Type of work
CD packaging

Specification

Description
--

5050466-4383-5-5

THOMAS RUSIAK | UNICORN

01. UNICORN | 3:31 | 02. UNICORN INSTRUMENTAL | 3:31 | 03. SPINNING VIDEO | 4:16 |
WRITTEN & ARRANGED BY THOMAS K. RUSIAK. CO. WRITTEN & CO. ARRANGED BY S.K. RUSIAK.

RECORDED AT THE ELEPHANT AND DIRTY SOUTH. MIXED
BY SEBROC AT COSMOS STUDIOS. MASTERED BY BJÖRN
ENGELMANN AT CUTTING ROOM. PRODUCED BY THOMAS
K. RUSIAK. PUBLISHED BY LED SONGS/UNIVERSAL MUSIC
PUBLISHING AB. COVER BY ZION GRAPHICS.

www.ledrecordings.com | www.ss6.com

++

+ Cuban Council

Client
Tigerbeat6 Records

Origin
California, USA

Title of work
Tigerbeat6 Records,
"Open up and say...@<%_|_^[[]]"

Art Direction/Design
Toke Nygaard, Michael Schmidt

Year produced
2003

Type of work
CD packaging

Specification
Use of material: Digital files, CD
Dimension: Standard CD digipak

Description
A fold-out cover designed for the
"Open up and say...@<%_|_^[[]]"
Compilation CD released by
Tigerbeat6 Records in San Francisco.
The design contents include numerous
nude pixel people and a dog.

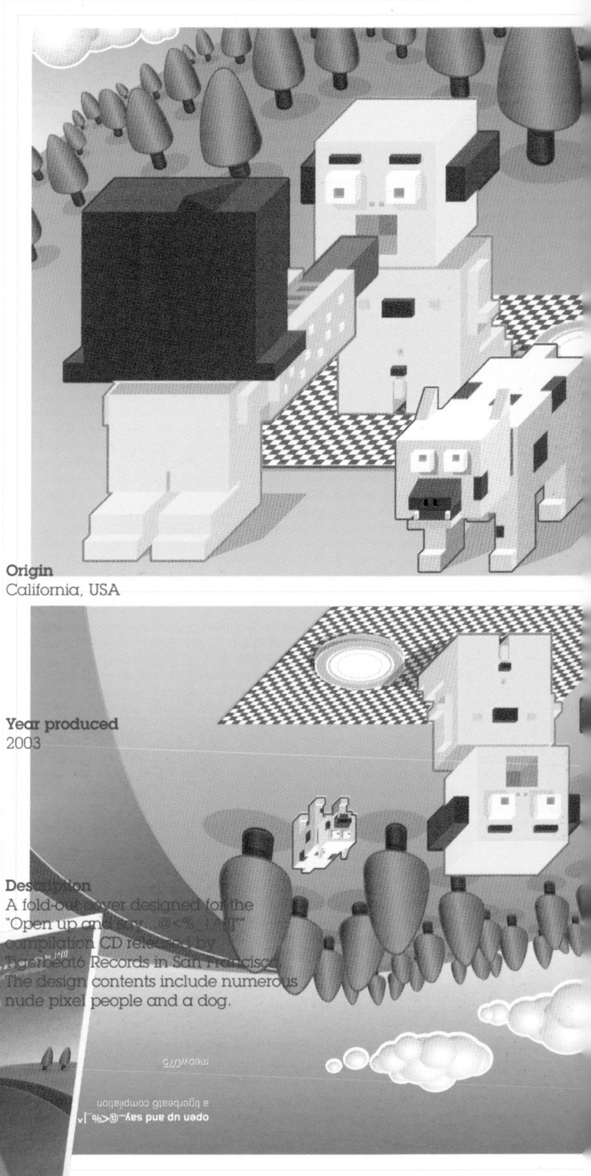

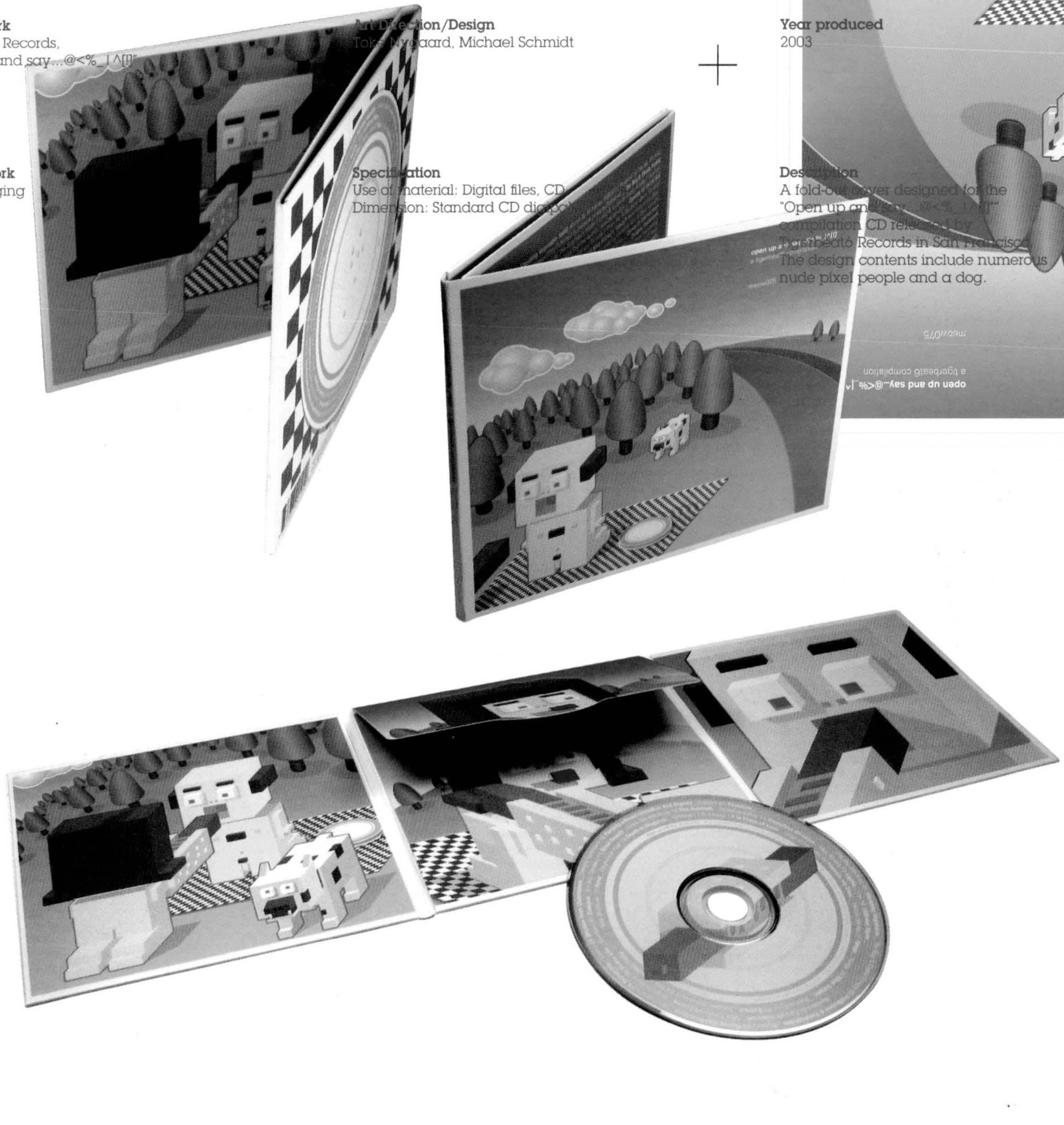

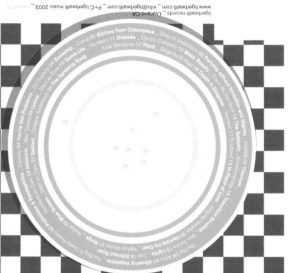

sleeve designed by
cubancouncil.com

tigerbeat6 records _ Oakland CA
www.tigerbeat6.com _ info@tigerbeat6.com _ P+C tigerbeat6 music 2003 _ meow075

+ Sagmeister Inc.

Client
Capitol Records

Origin
New York, USA

Title of work
OK Go's self titled debut album

Art Direction/Design
AD: Stefan Sagmeister
D: Matthias Ernstberger

Year produced
2002

Type of work
CD packaging

Specification
Dimension: 5 1/2" x 5" x 3/8"

Description
Ok Go is a young band from Chicago
playing sophisticated Pop. If their
music were a car, it'd be a 1980's
square Volvo.

6. SHORTLY BEFORE THE END (4:19)

How long did we all think this all would last?
Who could have counted days as they flew past?
But before we go, sing us a song.
Sing us a song to hum through the hours of dying.

Who would have thought it'd come as such a show?
A pink and silver day... who was to know?
Even as we go, sing us a song.
Sing us a song, to hum through the hours of dying.

7. RETURN (3:49)

Now it's years since your body went flat
And even memories of that are all thick and dull,
All gravel and glass.
But who needs them now,
Displaced they're easily more safe.
The worst of it now:
I can't remember your face.

Return.

For awhile, with the vertigo cured,
We were alive, we were pure.
The void took the shape of all that you were,
But years take their toll, and things get bent into shape.
Antiseptic and tired, I can't remember your face.

Return.

You were supposed to grow old,
Reckless, unfrightened, and old,
You were supposed to grow old.

Return. You were supposed to return.

8. THERE'S A FIRE (3:49)

Stop getting me off track.
I mean it, there's a problem here.
This time it is for real...
How can I make myself more clear?

I never say quite what I mean,
And never mean quite what I say,
And how did that get out of me,
And what the hell did I mean to say?

This time it is for real.
This is a real emergency.
This time I swear it is the truth...
This must be dealt with urgently.

I never say quite what I mean,
And never mean quite what I say,
And how did that get out of me,
And what the hell did I mean to say?

There's a fire. There's a fire.

I really mean it now.
This time I swear I have not lied.
This isn't like the last time...
I swear to God I have not lied.

I never say quite what I mean,
And never mean quite what I say,
And how did that get out of me,
And what the hell did I mean to say?

There's a fire. There's a fire.

1. GET OVER IT (3:16)

Lot of knots, lot of snags,
Lot of holes, lot of cracks, lot of crags.
Lot of naggin' old hags,
Lot of fools, lot of fool scum bags.
Oh it's such a drag, what a chore...
Oh your wounds are full of salt.
Everything's a stress and what's more,
Well it's all somebody's fault.

Hey! Get over it!

Makes you sick, makes you ill.
Makes you cheat, slipping change from the till.
Had it up to the gills...
Makes you cry while the milk still spills.
Ain't it just a bitch? What a pain.
Well it's all a crying shame.
What left to do but complain?
Better find someone to blame.

Hey! Get over it!

Got a job, got a life,
Got a four-door and a faithless wife.
Got those nice copper pipes,
Got an ex, got a room for the night.
Aren't you such a catch? What a prize,
Got a body like a bottle axe.
Love that perfect frown, honest eyes,
We ought to buy you a Cadillac.

Hey! Get over it!

2. DON'T ASK ME (2:46)

Quit acting so friendly.
Don't nod, don't laugh all nicely.
Don't think you'll upend me.
Don't sigh, don't sip your iced tea.
And don't say, "It's been a while..."
And don't flash that stupid smile.

Don't ask me how I've been.

Don't think I've forgotten,
You never liked that necklace.
So cordial, so rotten.

Kiss, kiss, let's meet for breakfast.
Don't show up so on time
And don't act like you're so kind.

Don't ask me how I've been.

Don't sit there and play just
So frank, so straight, so candid,
So thoughtful, so gracious,
So sound, so evenhanded.
Don't be so damn benign
And don't waste my blasted time.

Don't ask me how I've been.

3. YOU'RE SO DAMN HOT (2:36)

I saw you sliding out the bar.
I saw you slipping out the back door, baby.
Don't even try and find a line this time, it's fine.
Darling, you're still divine.

You don't love me at all.
But don't think that it bothers me at all.
You're a bad-hearted boy trap, baby doll, but you're...
You're so damn hot.

So now you're headed to your car.
You say it's dinner with your sister, sweetie.
But darling look at how you're dressed,
Your best suggests another kind of guest.

You don't love me at all,
But don't think that it bothers me at all.
You're a bad-hearted boy trap, baby doll, but you're...
You're so damn hot.

So who's this other guy you've got?
Which other rubes are riding hotshot, sugar?
I could have swore you said before, "No more, for sure."
What'd I believe you for?

You don't love me at all,
But don't think that it bothers me at all.
You're a bad-hearted boy trap, baby doll, but you're...
You're so damn hot.

Title of work
Mansun 'Kleptomania'
Harvey Dovey, Album & Single

Type of work
CD packaging

Client
+ Parlophone
++ Regal Recordings

Art Direction/Design
AD: Jeremy Plumb
D: Stuart Hardie

Specification
+ Use of material: Clear PVC outer
sleeve, 300gsm matt finish card
Dimension: 127 x 140 x 21 mm
++ Use of material: 300gsm Brown
Board, 300gsm uncoated white card
Dimension: CD digipak and 7"

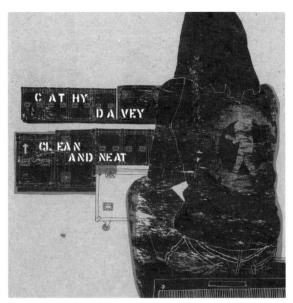

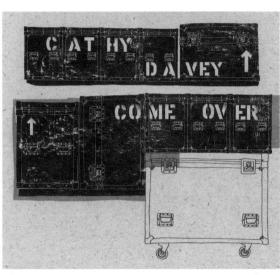

Origin
London, UK

Year produced
2004

Description
+ A standard digipak format was adapted to house three CD's and a 32-page booklet, all of which were encased in a clear PVC outer sleeve. The design used a typographic approach based around the word 'Kleptomania'; large hand painted letters were cut out and collaged together to produce a powerful design that flows throughout the whole package.
++ Cathy Davey's first EP 'Come Over' and subsequent single 'Clean & Neat' feature illustrations based around flight cases and boxes. Brown board was used to reflect the nature of the illustrations and give the packaging a tactile quality. This was then presented in a digipak format and printed on uncoated stock with a 12-page booklet glued to the inside face of the package.

++

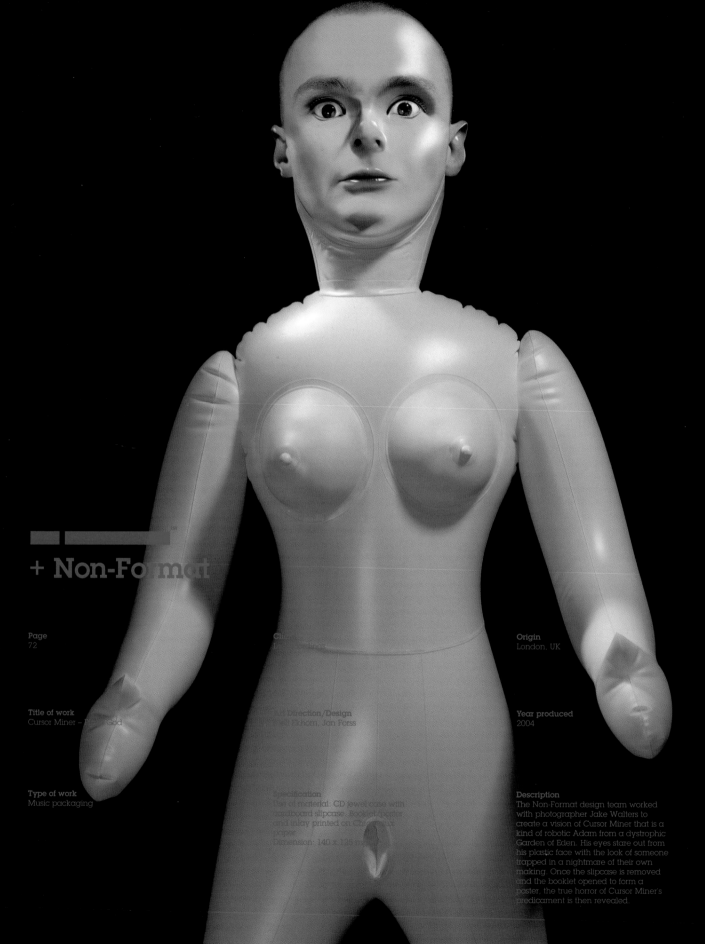

Page
72

Origin
London, UK

Title of work
Cursor Miner – Plug Me God

Art Direction/Design
Kjell Ekhorn, Jon Forss

Year produced
2004

Type of work
Music packaging

Specification
Use of material: CD jewel case with
cardboard slipcase. Booklet/poster
and inlay printed on Chromolux
paper
Dimension: 140 x 125 mm

Description
The Non-Format design team worked
with photographer Jake Walters to
create a vision of Cursor Miner that is a
kind of robotic Adam from a dystrophic
Garden of Eden. His eyes stare out from
his plastic face with the look of someone
trapped in a nightmare of their own
making. Once the slipcase is removed
and the booklet opened to form a
poster, the true horror of Cursor Miner's
predicament is then revealed.

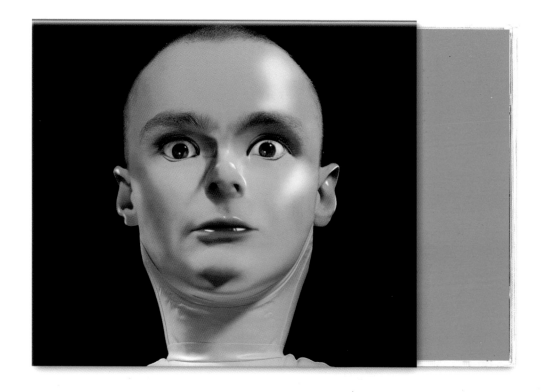

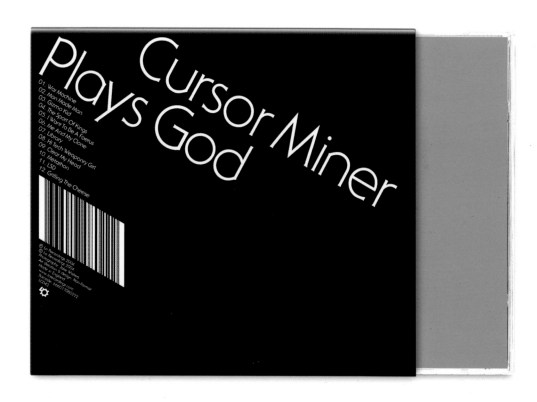

Cursor Miner
Plays God

01. War Machine
02. Man Made Man
03. Gizmo Kid
04. The Sport Of Kings
05. I Want To Be A Foetus
06. Me And My Clone
07. Library
08. Hi Tech Weaponry Girl
09. Clear My Head
10. Metrathon
11. LSD
12. Grilling The Cheese

℗ Lo Recordings 2004
© Lo Recordings 2004
An imaginary Jake Walters
Art direction & design - Non-Format
Made in England
www.lorecordings.com
Barcode
LCD06 5060173580772

+ Non-Format™

Page
74

Client
+ Accidental Records
++ Lo Recordings
+++ BMG Zomba

Origin
London, UK

Title of work
+ Mara Carlyle – The Lovely
++ King Of Woolworths – Re-diffusion
+++ Lo Editions

Art Direction/Design
Kjell Ekhorn, Jon Forss

Year produced
2004

Type of work
Music packaging

Specification
+ Use of material: CD packaging in
book bound form. 24-page section
printed on translucent paper
Dimension: 140 x 125 mm
++ Use of material: CD jewel case with
cardboard slipcase
Dimension: 140 x 125 mm
+++ Use of material: Three CD digipak
in a paper-cased slipcase box
Dimension: 140 x 125 x 30 mm

Description
+ This book bound CD packaging has
a cloth spine and an uncoated paper-
cased finish. The layering effect of the
illustrations of branches, roots, foliage
and insects is enhanced by the use
of a translucent paper similar to that
used in wedding albums.
++ The slipcase, printed with gold and
black typography slides off to reveal
a series of still life photographs. The
shots depict opulent place settings for
a series of meals. On closer inspection,
a rather unhealthy coating of mould
can be seen on the food suggesting a
down at heal existence.
+++ A set of three illustrations was created on the front of each digipak. The
three digipaks are all neatly housed
in a custom-made box. Contemporary
illustrations were commissioned for
each of the subsequent Lo Editions
box sets.

++

+++

Lo Editions

1.2.3.

Electric Sheep
Synthetic Pleasures
Bug Powder

Page
77

Client
Aesthetics

Origin
USA

Title of work
+ Daniel Givens – Age CD/LP
++ An Invitation To Play

Art Direction/Design
AD: Hans Seeger
+ P: Daniel Givens

Year produced
+ 2000
++ 2001

Type of work
+ Music packaging
++ Flyer

Specification
+ Use of material: Various
Dimension: Various
++ Use of material: Sponge
Dimension: Approximately 3"
in diameter

Description
+ Daniel is a musician who utilizes
organic influences (voice, strings,
horns etc) and culminates them
digitally in the studio. The fluid, yet
prickly organic/mechanic approach
is used. The matt finish jacket contains
the uncoated inner sleeves, aging as
the listener uncovers its layers.
++ A monthly invitation to one of
Chicago's longest – running music
series, Play.

PLAY

11/12 Miles Tilmann
11/19 Mr. Scruff
11/26 Synergy
12/03 One-F
12/10 Ambivert
12/17 Sõnik

At Danny's
1951 W. Dickens
Chicago

++

▪ ▪▬ ™
+ Heads Inc.

Page
78

Title of work
+ Bonaparte NY promo kit
++ Tomoko Yazawa / Transition
+++ Wanna buy a craprak?

Type of work
+ Promo kit
++ CD package

Client
+ Bonaparte NY
++ Monroe Street Music
+++ Carpark Records

Art Direction/Design
So Takahashi

Specification
+ Use of material: Museum board,
rubber band, letter-press
Dimension: 5 x 7
++ Use of material: CD jewel case/
silkscreen
Dimension: 143 x 125 mm
+++ Use of material: --
Dimension: 143 x 125 mm

Description
+ Bonaparte NY is a contemporary
wedding dress manufacturer. This
promo kit incorporates the traditional
wedding elements such as the cross
shape in the outer packaging. It is held
together by two rubber bands, similar to
two wedding bands. The type is letter-
pressed onto the cover. Each image
inside is on a loose sheet to facilitate
easy reference.
++ CD packaging for electronic music.
The cover changes when the first sheet of
the sleeve is slid out.
+++ The photographs are inspired by
the colours of the printing process, and
the process of layering is reflected in
the design, with the dimensions of the
photographs changing as the cover
sleeve is unfolded. A playful cover is
used to reflect the pop element of the
music.

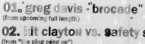

01. greg davis - "brocade"
(from upcoming full length)

02. kit clayton vs. safety scissors - "17-11"
(from "the ping pong ep")

03. marumari - "saka"
(from their out-of-print first cd "story of the heavens")

04. so takahashi - "blue, blue, electronic blue"
(exclusive track)

05. ogurusu norihide - "5:00"
(from "modern")

06. dinky - "no love"
(from "black cabaret")

07. freescha - "live and learn me"
(exclusive track)

08. casino versus japan "aquarium"
(from "whole numbers play the basics")

09. kid606 - "if my heart ever ran away
it would be looking for the day when right
beside you it could forever stay"
(from "the soccergirl ep")

10. takagi masakatsu "golden town with sunglasses"
(lusive track)

1. hrvatski - "equinox"
(clusive track)

2. signer - "interior dub"
m "low light dreams")

3. jake mandell - "beartrap!"
(clusive track)

eo: 242pilots "live at taklos"
akagi masakatsu "i'm computer, i'm singing a song"
arumari "way in the middle of the air"
jake mandell "the prince and the palm"

©2003 carpark records usa www.carparkrecords.com   desig : heads inc.
k 23 cd   www.   m.com
park p.o. box 20368 new york, ny 10009 usa   photo: julie m inic jacobsen
                                                                    masayo kishi

+++

+ Templin Blink Design

Client
Kunstuff

Origin
California, USA

Title of work
Archer Farms packaging

Art Direction/Design
AD: Joel Templin, Gaby Brink
D: Brian Gunderson

Year produced
2003

Type of work
Retail packaging

Specification
—

Description
A project to re-design the Archer Farms' identity and the packaging system for Target Stores. A modern leaf-like form and customized logo type is developed for the logo mark. The leaf shape then became the core element to create die-cut windows, shapes containing photography and a series of fun patterns. A family of animals are derived from the shape for the kids' food range.

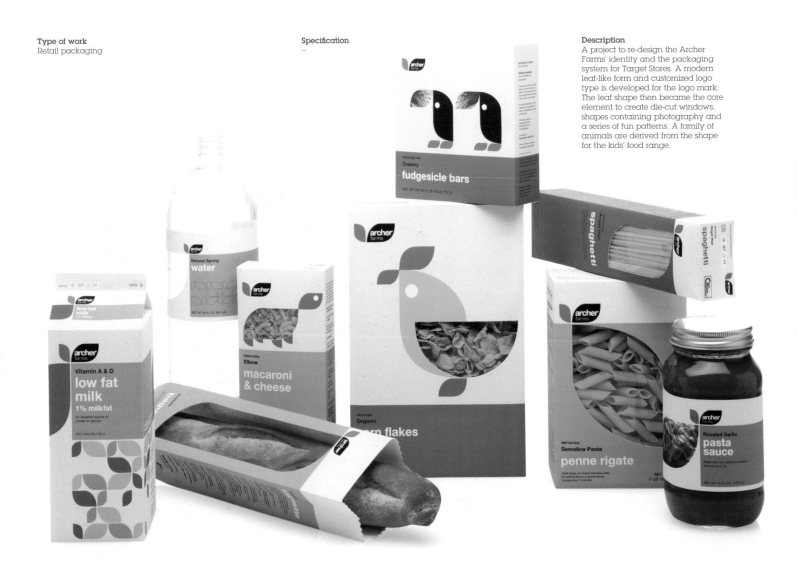

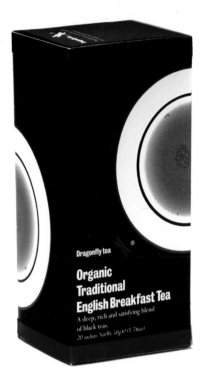

+ Pentagram™

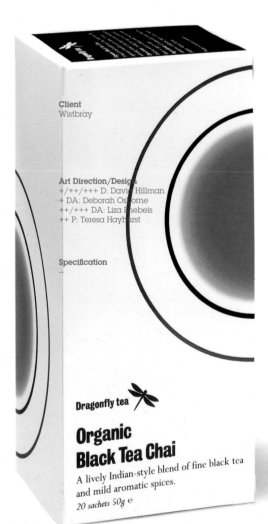

Title of work
+ Dragonfly Black range packaging
++ Dragonfly Teas identity and pack-
aging for Organic range
+++ Dragonfly Rooibos Naturally
Caffeine Free everyday tea range

Type of work
Packaging design

Client
Wistbray

Art Direction/Design
+/++/+++ D: David Hillman
+ DA: Deborah Osborne
++/+++ DA: Liza Enebeis
++ P: Teresa Hayhurst

Specification

Origin
London, UK

Year produced
+ 2004
++ 2000
+++ 2003

Description
Wistbray launched Dragonfly Tea, a
new range of premium organic teas
that introduced new, modern blends
to the UK market. This category has
traditionally witnessed a plethora
of colourful ethnic approaches to
packaging, making the shelf environment visually chaotic. A calm graphic
approach was used to show the quality and clarity of the products. The
project included three product ranges:
Dragonfly Organic, Dragonfly Rooibos
Everyday and Dragonfly Black Tea.

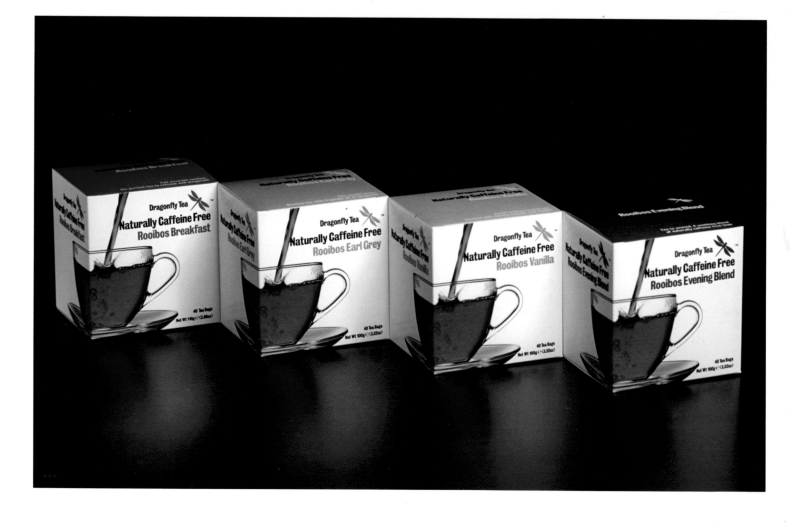

+ dossiercreative inc

Page
84

Client
dossiercreative inc.

Origin
Vancouver, Canada

Title of work
3XT

Art Direction/Design
AD: Don Chisholm,
D: Eena Kim

Year produced
2000

Type of work
Self promotion/client's Christmas gift

Specification
–

Description
The Christmas 2000 gift from dossiercreative inc. to their clients was a set of tea from three different regions. Each tea was packaged in individual tins decorated with motifs that were thematically linked to the associated regions. For example, the Chai tea's package incorporates Hindi script in its design. To convey the complexity of flavour of each tea, a rich layered design was created and the tactility of spot thermograph was added to accentuate the sensual quality of the experience.

mazorca
de maíz

La Sirena, una vez más apuesta por la calidad y el sabor de sus productos ultracongelados ofreciéndole dos Mazorcas de maíz, mazorcas hervidas, especialmente seleccionadas y casi listas para comer. Un toque exótico y sabroso para la cocina de hoy que aporta los hidratos necesarios para una dieta sana y equilibrada.

20-25 minutos   15-20 minutos   2 unidades

zanahoria
baby

La Sirena, una vez más apuesta por la calidad y el sabor de sus productos ultracongelados ofreciéndole Zanahorias baby, zanahorias redondeadas y tiernas especialmente seleccionadas casi listas para comer. Un toque de color, ligero y sabroso que hará de tus guisos o guarniciones el mejor acompañamiento.

9 minutos   5 minutos   9 minutos   300 gr.

MANZANILLA
ACEITE DE OLIVA
VIRGEN EXTRA

Carrefour

VERDIAL
ACEITE DE OLIVA
VIRGEN EXTRA

Carrefour

HOJIBLANCA
ACEITE DE OLIVA
VIRGEN EXTRA

Carrefour

™

+ Enric Aguilera

Page
87

Client
+ Carrefour
++/+++ La Sirena

Origin
Barcelona, Spain

Title of work
--

Art Direction/Design
+ AD: Enric Aguilera
D: Griselda Marti
++ AD: Enric Aguilera
D: Marcel Batlle, Griselda Martí
+++ AD: Enric Aguilera
D: Marius Zorrilla

Year produced
2004

Type of work
+Labels of oil for a supermarket chain
++/+++ Packaging for a chain of
frozen food

Specification
--

Description
+ These labels of oil suggest
different ways to use each type of
oil depending on its variety. All of
them aim to convey the healthy and
natural qualities through the use of
colours, images and typography.
++ A big project (more than thirty
references) of packaging frozen food
for La Sirena Company. The main
point was to show the quality and
freshness of their products. The elegant
and emphatic images incorporate
fully with the text.
+++ The text that describes the
product was printed with one-coloured
ink on a white background. These
elements were used to emphasize a
selection of basic products for daily
consume, which was as basic as their
prices.

BASIC
ARROZ TRES
DELICIAS

SALTEAR
8' Bolsa 1 kg

 la Sirena

BASIC
CANELONES
CON
BECHAMEL

UNIDADES MICROONDAS HORNO
6   5'   25' Caja 550 g

 La Sirena

BASIC
JUDIA PLANA
TROCEADA

MICROONDAS HERVIR OLLA PRESION
12'   12'   6' Bolsa 1 kg

 La Sirena

+++

+ loup.susanne wolf

Page
88

Title of work
+ Two Takeaway gift sets
++ Intercontinental gift sets

Type of work
Product conception and packaging

Client
Rösle GmbH & Co.KG

Art Direction/Design
+/++ CD: Silke Braun, Susanne Wolf
+ D: Saskia Bannasch

Year produced
2004

Specification
+ Use of material: PVC 300 mu
Dimension: 15 x 25 x 8cm, 15 x 35 x
8cm, 15 x 45 x 8 cm
++ Use of material: Polypropylene,
priplak Opaline 0,8 and fine
polystyrene
Dimension: 50 x 50 cm

Description
The RÖSLE Metallwarenfabrik is
the manufacturer of professional
kitchen tools. A strategic measure
is developed to open new market
opportunities to RÖSLE.
+ Two Takeaways chose trendy
subjects such as coffee, Mediterranean
cuisine and outdoor living/barbecue
to address to a young lifestyle-oriented
target group. The convincing and
innovative gift set with carrying
handle aims to create an impulse buy
for consumers. The sets are distributed
via the internet, in department stores
and furniture/ lifestyle shops.
++ Intercontinental reflects the
cultures of foreign countries, which
enrich the everyday life. The Bar Set,
Caribbean Nights and Cuban Rhythm
starts a world trip that goes via
Spain to Vino Tinto. Intercontinental
is addressed to a young, lifestyle-
oriented target group, which is living
the cosmopolitan way. Same as the
above, the sets are distributed via the
internet, in department stores and
furniture/lifestyle shops. In addition,
the sets are also distributed by
advertising medium companies and
used in the B2B area.

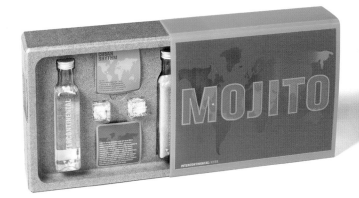

++

■■ ■■ ™

+ loup.susanne wolf

Page
90

Client
+/++ Kahla Thüringen Porzellan GmbH
+++/++++ The town council of Stuttgart

Title of work
+ Kahla Coffe Bar
++ Kahla Finger Food
+++ Investors event/Compass
++++ Investors event/Source of success

Art Direction/Design
+/++/++++ CD: Susanne Wolf
+ D: Sonja Linnemann
++ AD: Claudia Fuchs
+++ CD: Silke Braun
++++ AD: Erasmus Stillner, Susanne Wolf

Year produced
+/++/++++ 2003
+++ 2004

Type of work
+/++ Product conception and packaging
+++/++++ Packaging concept

Specification
+ Use of material: cardboard
Dimension: 40 x 40 cm
++ Use of material: cardboard
Dimension: 40 x 25 cm
+++ Use of material: cardboard
Dimension: 20 x 20 cm
++++ Use of material: cardboard
Dimension: 30 x 20 cm

Description
+/++ Kahla is developing into one of the most modern and innovative porcelain manufacturers in Germany. Kahla has entrusted loup.susanne wolf with both the conception and the packaging design of their high-value gift sets for the business-to-business products.
+++ Every year Lord Mayor of Stuttgart hosts an exclusive investors event in the state capital. An invitation is created for the international investors that conveys the characteristics of the event. In a box, there is a compass that shows the way into the state capital, acting as a personal signpost. Fluorescent colours are printed on the high-quality cardboard packaging, in which the sporting challenge lies. The glow in the night is analogous to the graphics of the needles and the four points of the compass. The navigation device is personalized by engraving the investor's name, thus assuring a valuable and lasting memory of the event.
++++ The City of Stuttgart — High-ranking international business executives are annually invited to an exclusive function in the state capital by the Chief Burgomaster of Stuttgart. His goal is to convince successful entrepreneurs of the merits in this location. Since Stuttgart has the second largest deposit of mineral water in Europe, loup.susanne wolf decided to create an invitation carrying the title, "The Source of Success".

+++

9.11° LÄNGE / 48.46° BREITE

++++

™

+ **Multistorey**

Page
92

Client
INK

Origin
London, UK

Title of work
Costa Del Salt and Puerto Del Pepper
packaging design

Art Direction/Design
AD: Harry Woodrow, Rhonda
Drakeford
D: Byron Parr

Year produced
2004

Type of work
Product packaging and branding

Specification
Use of material: Screenprinted board
Dimension: 90 x 90 x 235 mm

Description
Salt and Pepper grinders are
the shape of maracas, and the
packaging describes their dual use visually. The names Costa Del Salt and
Puerto Del Pepper were introduced to
describe the link between the grinders
and the Latin American origin of maracas. The boxes needed to provide
complete protection to the wooden
grinders, so they were designed as
closed boxes. The graphics applied to
them had to be descriptive and reflect the double meanings suggested
by the grinders. The stark black and
white illustrations were designed to
be instantly decipherable with witty
suggestions of what lies inside.

━━ ━━ ™

+ Kokokumaru

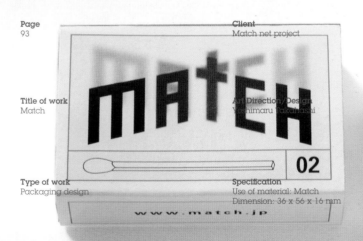

Page
93

Title of work
Match

Type of work
Packaging design

Client
Match net project

Art Direction/Design
Yoshimaru Takahashi

Specification
Use of material: Match
Dimension: 36 x 56 x 16 mm

Origin
Osaka, Japan

Year produced
2002

Description
–

Page
94

Client
Meiji Seika Kaisha, Ltd.

Origin
Tokyo, Japan

Title of work
100%ChocolateCafe

Art Direction/Design
groovisions

Year produced
2004

Type of work
Shop items

Specification
Base material: Plastic, paper, etc.
Dimensions: Many kinds

Description
Credit of 100% Chocolate Cafe:
Design produce – Koichi Ando
(Ando Gallery)
Art direction & Design –
Interior Design –

Client

+ Pentagram

Client
EAT

Origin
London, UK

Title of work
EAT

Art Direction/Design
D: Angus Hyland
DA: Charlie Smith, Sharon Hwang

Year produced
2002

Type of work
Packaging design

Specification
--

Description
The new brand identity for a London based premium sandwich bar chain highlights the honesty at the heart of their business. A new strap-line, The Real Food Company, was developed to further reinforce the brand value. The logotype employs bold sans serif, communicating the warmth and quality of the brand with clarity and a contemporary tone. This was complemented by a colour palette of warm, natural and brown hues. A range of vibrant minor colours was used for typography and detailing across food packaging, menu boards and other collateral.

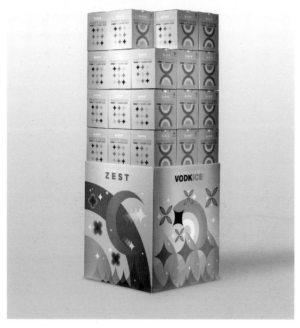

+ Époxy

Page
103

Title of work
Boomerang Cozmo campaign

Type of work
Advertising campaign

Client
Stéphane Duval/La Brasserie Labatt

Art Direction/Design
CD: Daniel Fortin, Georges Fok
AD: Marie-Hélène Trottier
D: Marie-Hélène Trottier
PdM: Hélène Joannette
CG: André Renaud
CS: Jean-Sébastien Ouellet
AE: Marie-Geneviève Parent

Specifications
--

Origin
Montreal, Canada

Year produced
--

Description
Boomerang Cozmo is the culmination of a full year of research and development during which product creators visited major markets around the world. Including Paris, London, New York and San Francisco, they examined the current consumer trends and tastes. Taste tests also showed that cranberry, which drives the success of the Cosmopolitan, is by far the most popular choice in all categories. Special attention was paid to balancing Boomerang Cozmo's sugar content and colour, which is pinkish leaning towards red.
The advertising campaign to support the Boomerang Cozmo launch built on the graphic environment developed for the Boomerang line – a greyish-silver background with animated geometric forms. In the case of Boomerang Cozmo, superimposed crescent moons in different variations of red were used due to the colour of the product.
Party Cube — packaging for alcoholic beverage that contains three different products of the same family, Vodkice - Vodka lime, Cozmo - Vodka cranberries and Zest - Ultra sharp lime flavour. Bold iconic graphics and dynamic colours were used to enhance its eye-catching brand statement.

Page
105

Title of work
Vodkice Packaging

Type of work
Advertising campaign

Client
Labatt Brewery Co

BOOMERANG
VODKICE

Art Direction/Design
CD: Daniel Fortin, George Fok
AD: Marie-Hélène Trottier
D: Marie-Hélène Trottier
PdM: Hélène Joannette
CG: André Renaud
AE: Marie-Geneviève Parent
BOISSON DE MALT AU GOÛT VODKA-LIME
MALT BEVERAGE WITH VODKA-LIME TASTE
Specification
341ml 7% alc./vol.

Origin
Montreal, Canada

Year produced
--

Description
It is a breath of air in an over-crowded advertising environment; the actual product is never seen. Impressionistic visuals and aural style quickly became associated with the product. The music was composed first and the visuals created with the music in mind. This explains why the visuals are so well integrated with the campaign. The whole mood was inspired by the music. There is a major restriction with the bottle format because there is a standard bottle for all the beers made by major breweries in Quebec, including Labatt. The bottles are all the same size, form and brown in colour. Since Vodkice is an alco-malt beverage (not a vodka-base drink), the standard brown bottle had to be used; so it was quite a challenge to give Vodkice a fresh look, innovative style and attitude.

+ Pentagram ™

Title of work
+ LZ 2001
++ Pago la Jara Anada and Dehesa
+++ Altos de Lanzaga

Type of work
+/+++ Labeling for wine bottles
++ Packaging design

Client
Compania de Vinos Telmo Rodriguez

Art Direction/Design
D: Fernando Gutiérrez
+/+++ I: Sean Mackaoui
++/+++ DA: Marc Catala

Specification
--

Origin
London, UK

Year produced
2002

Description
Telmo Rodriguez is an unique wine-maker who revives and re-creates Spanish wines that have slipped into obscurity. Labels and packaging were designed to reflect the unique, exclusive qualities of the brand, maintaining their high-quality profile and potential for success in an incredibly competitive marketplace. The Dehesa Gago (a Toro wine) label uses a lower case 'g' redolent of the profile of a bull - screen-printed onto the bottle to lend an earthy, tactile appearance. Labels for LZ and Altos de Lanzaga use variations of a molecular structure image by Sean Mackoui, a hand picking molecules can be seen as one picking a grape.

+

++

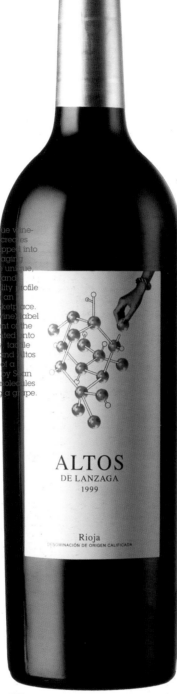

+++

Page
108

Title of work
OP Vodka

Type of work
Packaging design

Client
V&S Group

Art Direction/Design
Gabor Palotai

Specification
Use of material: Glass, paper
Dimension: 8 x 8 x 30 cm

Origin
Stockholm, Sweden

Year produced
2002

Description
OP Vodka, a new design for the traditional Swedish brand OP Aquavit created for the global market. The glass of the bottle is sandblasted in order to create an aura of coolness. The innovation in design transformed the transparent tradition of the 1920s' exclusive Swedish grace to a sophisticated ultra modern Swedish design. The magnificent slim glass of the OP Vodka bottle holds onto tradition while the simple logo and graphics bring the design up to date. The cool aura is confirmed by the sleek appeal and sensually harsh feel of the sandblasted bottle. Typed in the Mittelschift – a font created for the German Autobahn – the letters are clear, concise and easily recognizable. At the top of the name OP is a pair of arrows, a symbol seen across the continents meaning push, do, or proceed. OP Vodka want the consumers to participate and push the limits of the fine spirits to experience the high sensuality of being extraordinary cool.

+ Enric Aguilera

Client
+ Carrefour
++/+++ Caves Torelló

Origin
Barcelona, Spain

Title of work
+ Vites Virides
++ Gran Torelló
+++ Marc Torelló

Art Direction/Design
+ AD: Enric Aguilera
D: Marcel Batlle
++/+++ AD: Enric Aguilera
D: Marius Zorrilla

Year produced
+/++ 2004
+++ 2003

Type of work
+ Labels of wine for a supermarkets chain
++ Label of cava (type of cava produced in Catalonia)
+++ Label for a marc of cava (alcoholic drink that comes from a special distillation)

Specification
--

Description
+ Simplicity and elegance are the key elements that the client wants to emphasize on for their ecologic products. The only difference between each label is just the image: A vine leave for the youngest wine and an oak section for reserve quality wine.
++ Gran Torellós logo becomes the main element of this project. Through the absence of graphic and chromatic elements, it shows a clear idea of what this Cava represents: Superior Quality.
+++ High Quality product with a very limited production was the base of this project. With an austere label, the prominence of this pack is the neck part where through images of a stump, it shows us the singularity and precedence of this product.

++

+

+++

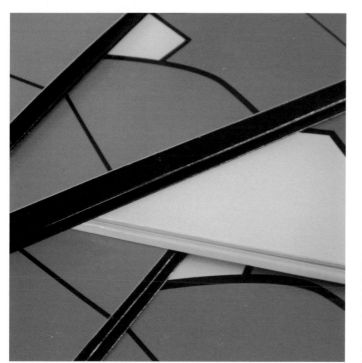

Page
111

Client
+ Marine Parade
++ Geisha

Origin
Nottingham, UK

Title of work
+ Evil Nine single & album
++ Geisha matchbox

Art Direction/Design
+ Rob Coke
++ AD/D: Dan Moore
D: Lydia Lapinski

Year produced
2004

Type of work
+ Vinyl packaging
++ Matchbox

Specification
+ Use of material: –
Dimension: 300 x 300 x 8 mm
++ Use of material: custom matchbox
and coloured matches
Dimension: 56 x 47 x 6 mm

Description
+ The sleeve and campaign for
Evil Nine's LP, 'You Can Be Special
Too' feature illustrations of the band
based around a sketch by the artists
themselves. As well as the sturdy new
super jewel case format, the album is
also available in a luxurious double
gatefold vinyl, which includes stickers
and masks of the characters, allowing
you to 'be evil' while listening!
++ Geisha is a Far East-inspired
restaurant, bar and nightclub in
Nottingham, UK. The core identity is a
marque inspired by ancient Japanese

Kamon, the family crests displayed on
hangings outside the home or place of
business. The plum blossom or 'ume'
is associated with spring, new life,
and more occasionally with Geishas.
The symbol is designed in both a
standard weight and a restricted size
version with counters removed. The
understated theme is continued with a
simple and elegant logotype featuring
a custom ligature.
The matchboxes exude exclusivity,
containing custom-made black
matchsticks with cherry blossom
pink strikes. The tone of the project is
unashamedly elitist and resolutely
aspirating with the slogan: 'Geisha is
for those who know their sake from
their saketini.'

PG.114 > 163
TOUCH

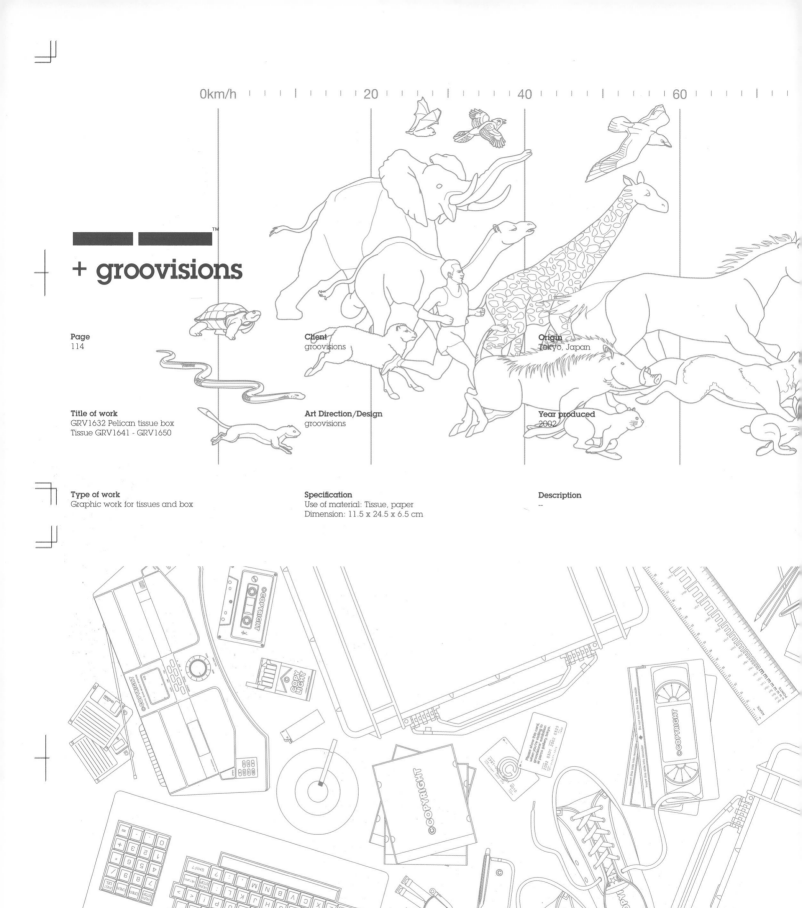

+ groovisions

Page
114

Title of work
GRV1632 Pelican tissue box
Tissue GRV1641 - GRV1650

Type of work
Graphic work for tissues and box

Client
groovisions

Art Direction/Design
groovisions

Specification
Use of material: Tissue, paper
Dimension: 11.5 x 24.5 x 6.5 cm

Origin
Tokyo, Japan

Year produced
2002

Description
--

+ groovisions

Page
117

Client
BNN Co., Ltd.

Origin
Tokyo, Japan

Title of work
GRV2000

Art Direction/Design
groovisions

Year produced
2000

Type of work
Work collection

Specification
Use of material: Cardboard, plastic and iron
Dimension: 10 x 23 x 25 cm

Description
--

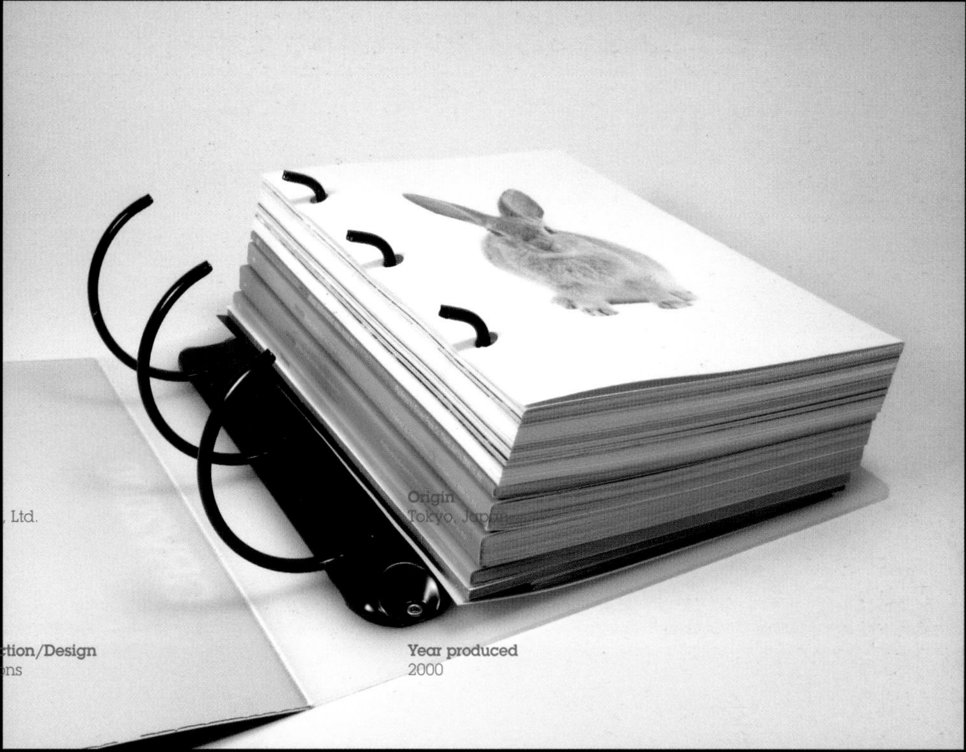

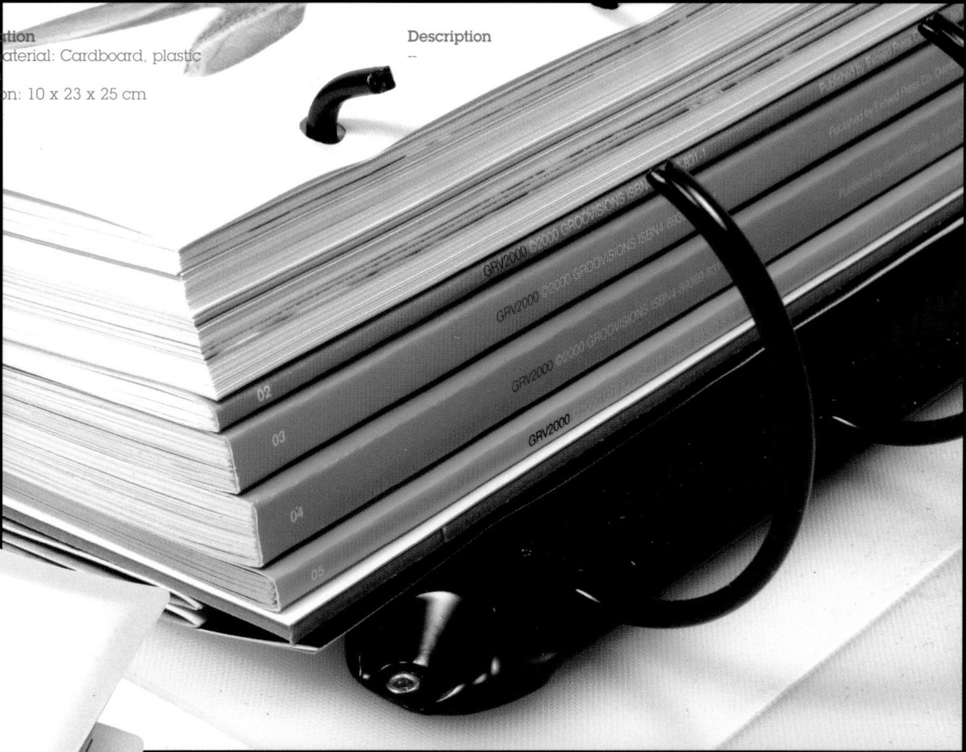
GRV2000 ©2000 GROOVISIONS
GRV2000 ©2000 GROOVISIONS
GRV2000 ©2000 GROOVISIONS
GRV2000
02
03
04
05

GRV2000
©2000 GROOVISIONS
ISBN4-89369-831-1
Published by Exceed Press Co.
Distributed by BNN co.,ltd
BNN

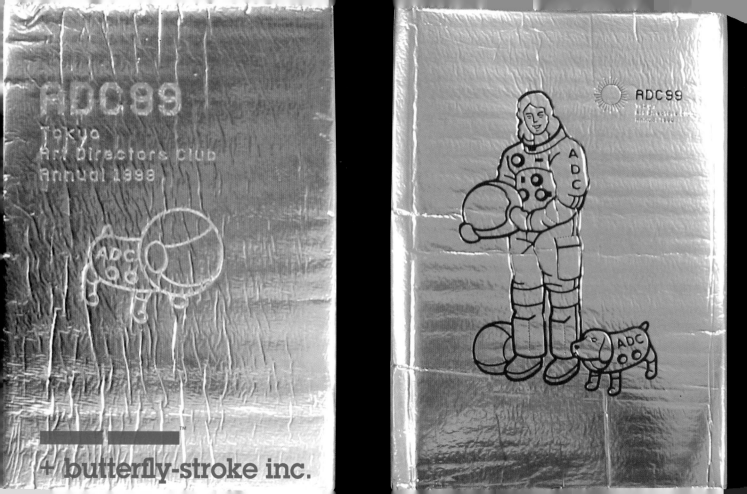

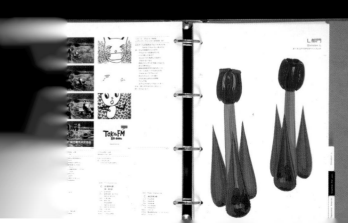

+Sartoria Comunicazione

Title of work
CUBE 01 & 02

Type of work
Limited edition books

Client
Sartoria

Art Direction/Design
AD/D: Giorgio de Mitri
D: Gilles Cenazandotti

Specification
Use of material: Printed cardboard,
foam, paper
Dimensions: Cube 01 – 27 x 27 cm;
Cube 02 – 29 x 29 x 39 cm

Origin
Modena, Italy

Year produced
Cube 01 (spaceship) 2000, Cube 02
(book stand) 2002

Description
Cube is an in-house project that
began in 1999, and the theme of
the project is 'no commercial traffic
in Cube'. Cube is the culmination of
Creative Director Giorgio de Mitri's
travels, serendipitous encounters and
a lifelong passion for provocative,
meaningful and rare images.
Gilles Cenazandotti designed the
cardboard package for Cube 01,
which contained a copy of Cube,
a DVD, and a limited edition long-
sleeve T-shirt. Once assembled, the
package became a spaceship or a
small table to rest Cube upon. Giorgio
de Mitri designed the special edition
for Cube 02, where solid black foam
was added to both the front and back
cover to create a kind of elevated
bookstand. Here the idea evolved
from a separate bookstand (Cube 01)

Page
22

Title of work
Carne Fres

Type of work
Annual report

Client
Architecture University of Rome
(Università di Roma Tre)

Art Direction/Design
Enrico Bonafede

Specification

Year produced
2003/4

Description
Carne Fresca [Fresh Meat] is a
student annual report 2003/2004
of Architecture University of Rome
[Università di Roma Tre]. The
University is situated in an ex-
slaughterhouse. Carne Fresca means
"new figures in the future architecture
scenery", which is similar to the term,
"Fresh Meat".

™

+ Brandia

Page
124

Client
CTT Correios

Origin
Lisboa, Portugal

Title of work
Experimenta/CTT Correios/Brandia
Organic Stamp

Art Direction/Design
Goncalo Cabral

Year produced
2003

Type of work
Packaging/graphic design

Specification
Dimension: 190 x 135 x 10 mm

Description
Organic Stamp is a project made for
CTT Correios de Portugal (national
mail) to be part of the Experimenta
Design Bienal de Lisboa in 2003

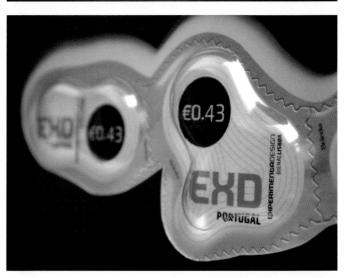

+ Ollystudio

Client
Levi's Vintage Clothing

Origin
London, UK

Title of work
LVC Dream Land Dwell

Art Direction/Design
AD/D: Oliver Walker
P: Kent Baker

Year produced
2002

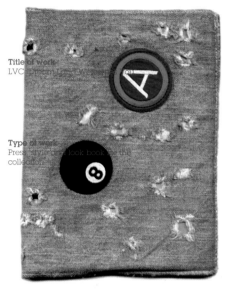

Type of work
Press style and look book for the collection

Specification
Use of materials: Denim, Cairn board, Book binding Tape and 150 gsm silk paper
Dimension: 180 x 230

Description
As the only piece of promotion for the LVC collection, this book had to carry product information, press information and style-led images that clearly showed the product. The finished product was a limited edition case bound section-sewn book with a loose denim dust jacket. The overall feel was to create a book that had the same characteristics as the jeans. It was created from tough simple materials then distressed so that each book had its own individuality. The photo shoot was done as a travelogue where the destination was unimportant, while what they encountered on the way formed the backdrop for the images. The images were then laid out on pages as if the book had been put together as the book was finished off with the jacket, which had stitches sewn onto it from the collection.

213 xx (type I) - heavy used

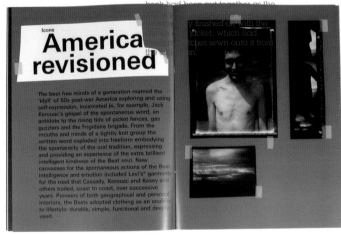

Icons
America revisioned

The best free minds of a generation roamed the 'idyll' of 50s post-war America exploring and using self-expression, incarnated in, for example, Jack Kerouac's gospel of the spontaneous word, an antidote to the rising tide of picket fences, gas guzzlers and the Frigidaire brigade. From the mouths and minds of a tightly knit group the written word exploded into freeform embodying the spontaneity of the oral tradition, expressing and providing an experience of the extra brilliant intelligent kindness of the Beat soul. New canvasses for the spontaneous actions of the Beat intelligence and emotion included Levi's® garments for the road that Cassady, Kerouac and Kesey and others trailed, coast to coast, over successive years. Pioneers of both geographical and personal interiors, the Beats adopted clothing as an enabler to lifestyle: durable, simple, functional and deeply used.

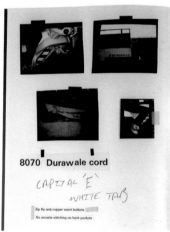

8070 Durawale cord

CAPITAL 'E'
WHITE TAB

Zip fly and copper waist buttons
No arcuate stitching on back pockets

CAPITAL 'E'
RED TAB

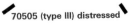

70505 (type III) distressed

+ Helena Fruehauf

Page
126

Client
Helen Fruehauf

Origin
New York, U.S.

Title of work
Halfbreed02*

Art Direction/Design
Helen Fruehauf

Year produced
2002

Type of work
Book

Specification
Use of material: Inside: Light blue uncoated text paper, vellum finish, 80#/cover 100#; Ultra white uncoated text paper, vellum finish 80#; Ultra white cast-coated paper, glossy 80#; Ultra white Poly and Silk fabric/Cover: Silk Fabric/Production: Single pages, French folds and fold outs; Japanese Binding
Dimension: 18.5 x 25.5 x 3.5 cm (7 1/4" x 10" x 1.5")

Description
Halfbreed02* is an attempt to create the designer's "own" authentic visual language based on her German and Chinese backgrounds. These rich cultures of ornamentation serve as a starting point of her investigation. The two heritages merge through various processes of hybridization, touch aspects of traditional theories and practices, as well as the contemporary vernacular. The journey concludes with a proposal for the application of ornamentation today, and it introduces a methodology that personalizes and transforms into an individual and possibly authentic language.

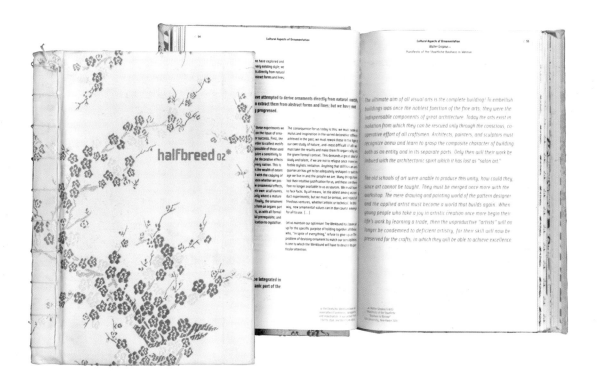

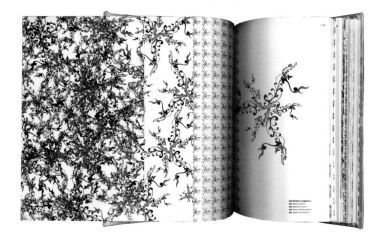
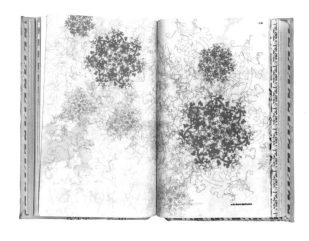

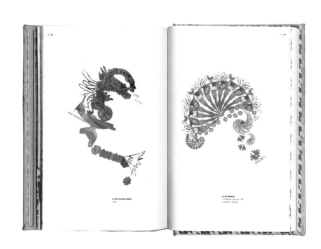

™

+ additional

Page
128

Client
additional

Origin
Italy

Title of work
additional – permeable packaging

Art Direction/Design
Rüdiger Schlömer

Year produced
2004

Type of work
Packaging design

Specification
Use of material: Wool, self-adhesive
envelopes, plastic bags, basic
packaging
Dimension: Multi-scale

Description
additional is a decentralized dispersal
form of packaging, attaching itself
to existing product dynamics. a
"Permeable packaging" allows a
fuzzy diffusion strategy. So far this has
been distributed in self-adhesive sets
which are stuck to a diverse range of
carriers. The concept of permeable
packaging widens possibility by
spreading out across the carrier.
Recent developments such as
additional smart dust (consisting of
wool lints) and viral pompons open
up a whole new area for urban micro-
invention.

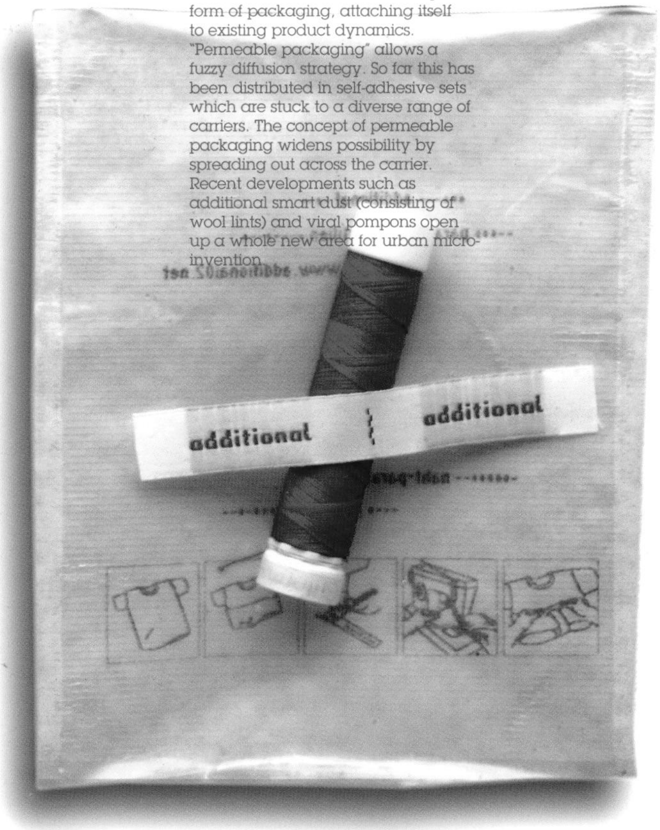

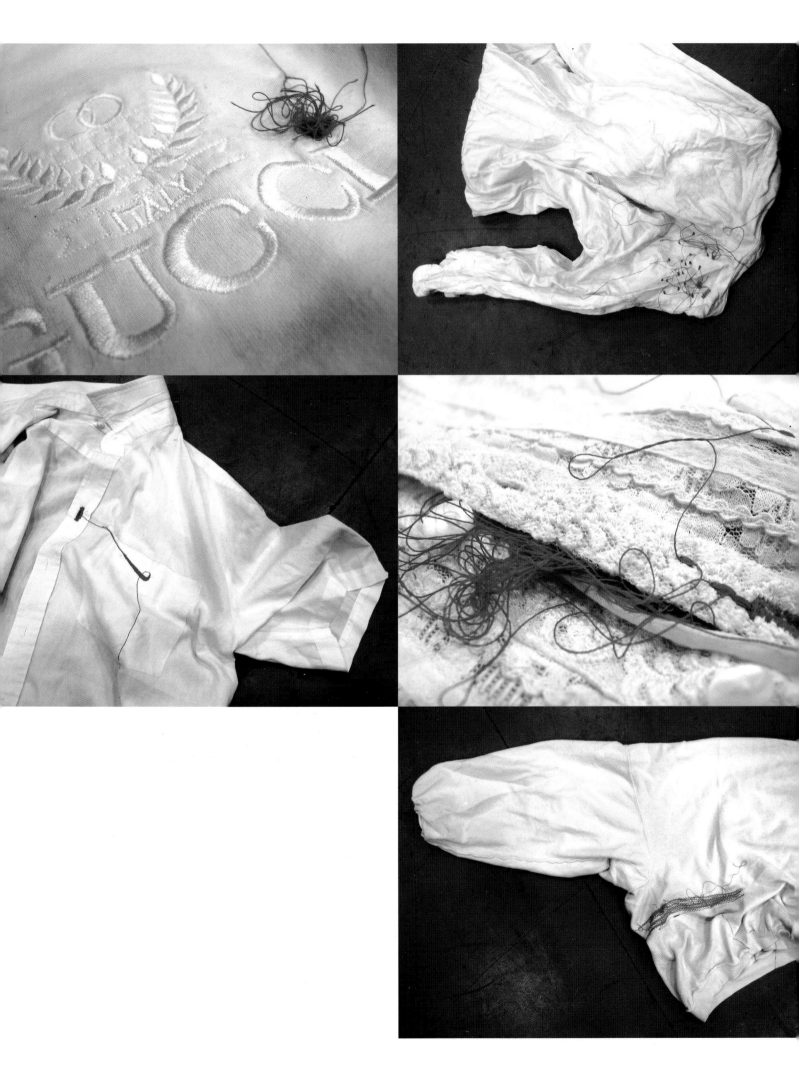

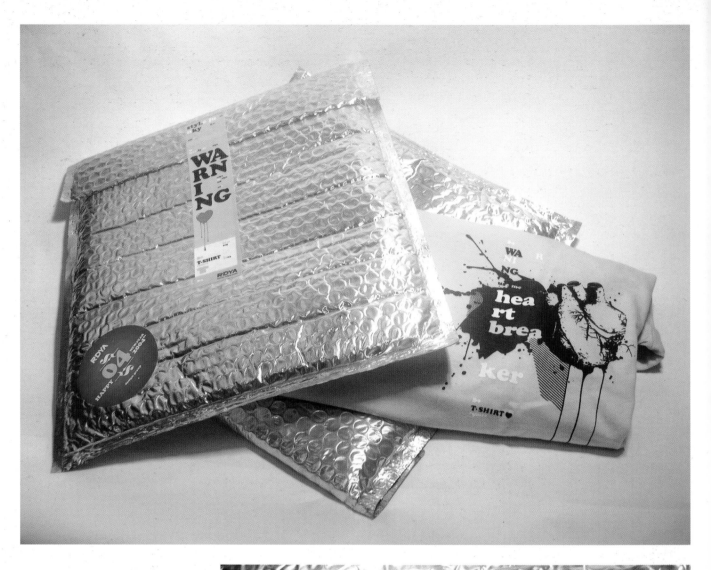

▬▬▬ ▬ ▬™

+ RDYA

Page
130

Client
RDYA

Origin
Buenos Aires

Title of work
RDYA T-shirt packaging

Art Direction/Design
Ricardo Drab

Year produced
2003-04

Type of work
T-shirt/tin packaging/silvery bag

Specification
Use of material: Tin/silvery cloth
to make the bags

Description
RDYA T-shirts were given to their clients as New Year presents. A diverse collection was elaborated to attract the public's attention. A tin was used as the packaging and merchandising tool, which was eye-catching enough in a retail setting. Another option for the present was to make a bag for the T-shirt with a special cloth in silvery colour.

+ Form™

Title of work
UniForm Safe-T mailer

Type of work
Fold out poster and T-shirt packaged
in plastic air pack

Art Direction/Design
AD/D: Paula Benson, Paul West
D: Claire Warner

Specification
Use of material: Poster – litho printed
on paper; Air-pack – plastic blow up
air bag
Dimension: 250mm x 180mm
(air pack)

Year produced
2003

Description
This mailer was created as a special
pack to launch the UniForm Safe-T
range, which was influenced by
safety graphics and information. A
fold-out poster detailing the range
and promotional T-shirts were placed
inside the air pack which was then
pumped up with air. These were sent
to key press contacts and bands as
presents. The address and stamp were
attached directly to the outside and
sent by normal post.

+ Kokokumaru

Page
132

Client
Takeo paper Co. ltd

Origin
Osaka, Japan

Title of work
Dust 100

Art Direction/Design
Yoshimaru Takahashi

Year produced
1998

Type of work
Book packaging

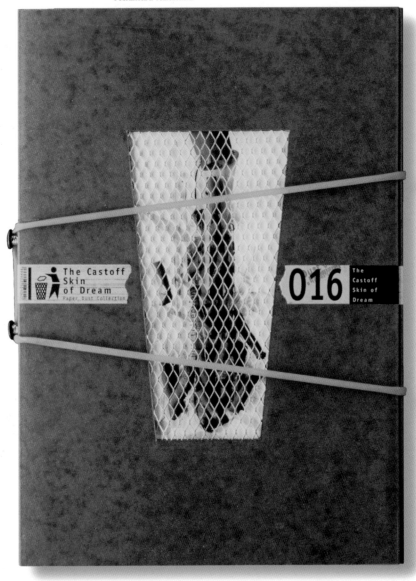

+ Less Rain

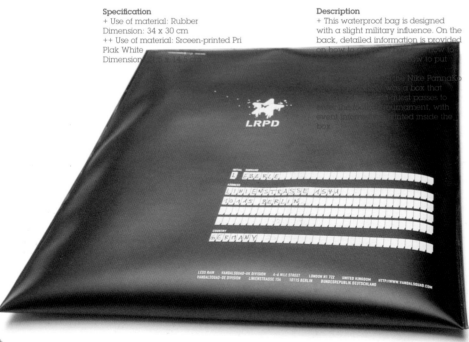

Title of work
+ Less Rain Vandal Squad T-shirt
Package
++ Nike PannaKo Invitation Package

Art Direction/Design
+ AD/D: Lars Eberle, D: Ossi Wising
++ AD/D: Lars Eberle

Year produced
+ 2004

Type of work
+ Packaging design
++ Event promotion

Specification
+ Use of material: Rubber
Dimension: 34 x 30 cm
++ Use of material: Srceen-printed Pri
Plak White
Dimension:

Description
+ This waterproof bag is designed
with a slight military influence. On the
back, detailed information is provided
on how to ... how to
put ...

... the Nike PannaKo
... was a box that
... guest passes to
... the football tournament, with
event information printed inside the
box

™

+ Sartoria Comunicazione

Page
136

Client
Nike

Origin
Modena, Italy

Title of work
White Dunk

Art Direction/Design
CD: Giorgio de Mitri
D: Sartoria Comunicazione.

Year produced
2003

Type of work
Invitation for the White Dunk
exhibition at Palais de Tokyo

Specification
Use of material: White leather, white
stitching, printed cardboard

Description
White Dunk at Palais de Tokyo was
an exhibition launched during
the Paris fashion week. Nike had
previously challenged twenty-six
Japanese designers, animators, mold
makers and artisans to create original
artworks based on the Nike all-White
high top basketball Dunk. The VIP
invitation to the White Dunk exhibition
came in the form of a square leather
zip case embroidered with the White
Dunk symbol. It contained a deck of
hard double-faced cardboard cards
with photographs of the artwork by
Harri Peccinotti and on the opposite
side, a brief biography of the artist.
The complete deck of cards was used
as the invitation and catalog all in
one.

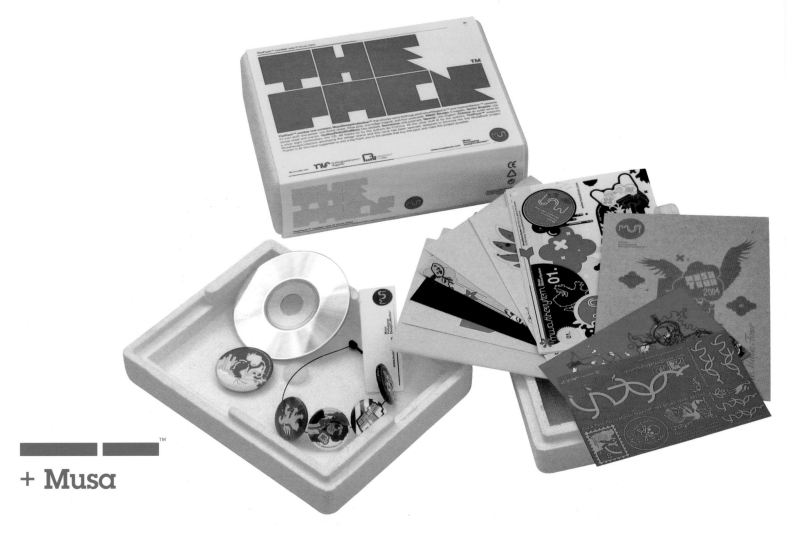

+ Musa

Page
138

Client
Musa Tour 2004

Origin
Lisboa, Portugal

Title of work
The Pack 01

Art Direction/Design
CD: Musa (Paulo Lima, Ricardo Alexandre, Raquel Viana)
The Pack Concept: Musa (Paulo Lima, Ricardo Alexandre, Raquel Viana); Contents design: Musa, RMAC, Vector Brigade, YouNeedToColour, Secretonix; Sound design: Crónica in first Musabbok CD

Year produced
2004

Type of work
Packaging

Specification
Use of materials: Expanded Plysth colour printe plastic film
Dimension, 18

Description
The Pack 01 is a free concept pack focusing attention on the enclosed material. Its purpose is obvious, the name is simple and shape regular. The challenge was to make a recognizable brand by using uncommon material while being inexpensive, light weight and resistant for effective mail delivery. The Pack is the innocuous box that encloses exciting products.

+ Milkxhake

Page
139

Title of work
"Shoot Better!" —
Re-design of 135-film roll

Type of work
Contribution

Client
Cream Magazine, Hong Kong

Art Direction/Design
Milkxhake

Specification
—

Origin
Hong Kong

Year produced
2003

Description
Cream Magazine, an alternative pop culture magazine published in Hong Kong, produced a special feature about 35mm films in November issue 2003. Milkxhake was invited to re-design a 135 film-roll, and "Shoot Better!" was developed to act as a "simple reminder" before pressing the shutter. The slogan was: "Shoot Better, keep an eye on every-thing." "Shoot Better, keep an eye on everywhere." "Shoot Better, keep an eye on everyone."

+ Airside

™

Page
140

Client
Sony Creative Products (SCP), Japan

Title of work
Dot Com Refugees

Art Direction/Design
Airside

Type of work
Toys, pencil cases, mugs

Specification
Use of material: Moulded plastic
Dimension: Various

Origin
London, UK.

Year produced
2003

Description
These people were working at a big Dot Com company in London, but became jobless when the company went bust. They are all looking for jobs now – maybe you can help?

RED SNAPPER REMIXED BY:

BLUE STATES
BROADWAY PROJECT
THE CREATION
DEPTH CHARGE
EUROPA 51
KNOWLEDGE OF BUGS
ODD MAN
PEDRO
RADIOACTIVE MAN
ROTHKO
RICH THAIR
SUSUMU YOKOTA

REVERSE SIDE STICKER FOR ITEM:

RED SNAPPER
REDONE

BARCODE: 6 66017 06972 2
LCD39

HTTP://WWW.LORECORDINGS.COM
MADE IN ENGLAND

Client
Lo Recordings

Art Direction/Design
Kjell Ekhorn, Jon Forss

Specification
Use of material: CD jewel case with
woven cloth label and printed sticker
Dimension: 140 x 125 mm

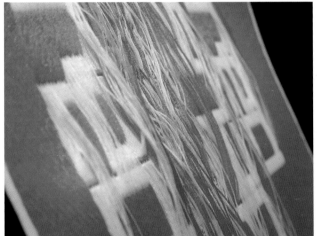
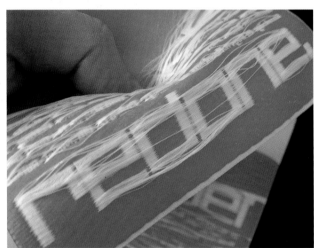

Origin
London, UK

Year produced
2003

Description
This music packaging for Red Snapper
has a woven label inserted onto the
inner tray of a jewel case. The label
has been designed so that the reverse
side, which has loose strands of red
and white cotton, is used to display
the artist's name and title in white
letters on a red background. The
'good' side of the label has the title
displayed back to front in red letters
on a white background and can be
seen beneath the CD along with the
full track listing.

Page
145

Title of
+ Lem
++ Rep
and So

Type of work
+ Mixed media
++ Limited edition 7" singles

Specification
+ Use of material: Plastic, card, lead,
wood, rubber, metal...
Dimension Various
++ Use of material: Denim and
Sackcloth. Screen-printing. Picture
Disk, Condom (in back pocket of
denim sleeve)
Dimension: 7 inches

Description
+ The small (approx 8") plastic bag
was given out to guests at Lemon
Jelly's Patagonia tour in 2003. The
contents included a bingo card
along with a pencil (aka 'Lemon
Jelly Doodle Machine' – etched onto
the pencil) for playing bingo, a red
balloon which was blown up and
thrown around the auditorium, an 'I
love Lemon Jelly' badge and various
postcards.
++ These 7" bootlegs were hand
made – one in Denim and the other
in Sackcloth. The denim one has
a condom in the pocket and the
Sackcloth one has a gold picture disk.

++

+ **Meem**

Client
Meem

Art Direction/Design
Michael Moebus

Type of work
CD packaging

Specification
+ Use of material: Screen-printed cotton, felt, button (sewn together)
Dimension: 140 x 140 mm
++ Use of material: Stained and unstained three-ply wood (front inside nia burnt into wood with hot stamp – rear & inside ink stamp), felt, velcr printed card
Dimension: 135 x 135 mm

Origin
Redfern, Australia

Year produced
+ 2003

Description
+ Two long pieces of cream-coloured cotton are screen-printed with brown ink and then sewn together around the edges with brown cotton thread to make a front and a back. This newly sewn piece is then wrapped around two felt squares and sewn along the sides.
The manufacture of the product was undertaken by various factory outlets throughout Sydney, Australia.
++ Three squares of three-ply wood house the CD. The two outside pieces are stained, the middle piece is unstained and has a large circle cut to fit the CD. (The middle piece of wood also has four smaller circles, one cut into each corner. This was a design error – as the design for the product constantly changed throughout its production).
On the front panel is a hot stamp insignia, which has been burnt into the wood. The back panel has a small ink stamp in the bottom right corner, and two velcro circles in each corner of the left hand side. A piece of brown felt is glued between the wood sections, which creates a spine (like a book), and this felt piece then extends past the front section (in length). This added length of felt has a smaller thinner piece of wood glued to it and has velcro circles stuck to each end. The velcro circles align and therefore create a way of closing the case.
This smaller piece of wood has an ink stamp (with text), and is individually hand-numbered. Finally, a printed circular piece of card is glued to the inside, containing track information and other details.
This project was undertaken by Meem over a five-month period between 1999 and 2000, with four hundred and four copies handmade individually by Meem.

+ Ollystudio

Page
148

Client
+ Universal Music Publishing
++ Alexander McQueen

Origin
London, UK

Title of work
+ William Orbit Promotional CD
++ Alexander McQueen 10
New York shop opening invitation
and limited edition CD packaging

Art Direction/Design
Oliver Walker

Year produced
2002

Type of work
CD packaging

Specification
+ Use of material: Clear acrylic and
special yellow glow edge acrylic discs
Dimension: 120 mm Diametre
++ Use of material: Acrylic, all black
CD's, black board, cartridge paper.
Dimension: 130 x 130 mm

Description
+ This special limited edition packaging was produced to promote the artist as a composer and producer within the film and advertising industry. The three-CD set was held in between the two screw-together acrylic discs. Two of the CD's contained William's selected tracks, whilst the other had track listings and legal information printed on both sides. The two music discs were produced on yellow glow edge acrylic discs, and only one line of type was used on the surface of the discs. This gave the owner the option of having either a yellow face or a silver face showing through the clear acrylic outer discs. When viewed from the side, the glowing edges of the disks can still be seen.
++ The invitation for the shop opening had ten tracks from the McQueen fashion shows, mixed by John Gosling. The main identity used on the disc and pochette was an illustration based on an image of a model taken from a fashion show. The materials used were special boards, papers and plastics that had black centres. This ensured that there were no white surfaces or edges anywhere on the pack. The special quality of the packaging was finished off by screen-printing the colour pink. The Perspex screw-together pack was produced and sold in the New York shop. The purpose of the packaging is purely to create an extravagant CD packaging to complement the McQueen collection.

™

+ Gunnar Thor Vilhjalmsson

Page
149

Title of work
Auxpan – Elvar EP
(Artist: Auxpan/Thor Jensen)

Type of work
CD cover

Specification
Use of material: Cardboard,
Vlieseline and Velcro
Dimension: 16 x 14 cm

Art Direction/Design
Gunnar Thor Vilhjalmsson

Year produced
2003

Description
Auxpan is a noise music artist living in Reykjavik. He makes noise with his home-made equipment, puzzled together from broken radios, greeting card modules, equalizers and everything he can get sound out of or alter sounds with.
The micro-label, Vasadisko (founded by Gunnar Thor Vilhjalmsson as a BA graduation project), released his six-track EP called Elvar in a limited edition of twenty copies.
The CD cover was made of cardboard. A sticker with printed graphics was laid over this cardboard, and enclosed with a white and red Velcro. The cover came in a vlieseline pocket with a silk-printed flower pattern, which was used as the main graphic element of the cover.
Auxpan uses sine waves in his work, and to emphasize this, a special feature was added to the package.

Whenever the cover was opened, an interactive sound module from a greeting card would welcome you with the sound of a sine wave that lasted for about ten seconds. This was also used as the sixth track of the EP. Inside, there were stickers, one with the Auxpan logo and the other with Vasadisko logo, and a booklet.
A personal studio portrait of Auxpan was used because the EP was named after him, Elvar. Auxpan turns hymns and noise of machines into an attractive mixture, so the graphics used were a metaphor for that, flowing from raw schematics to beautiful flower patterns.

+ Sartoria Comunicazione

Page
150

Client
+ Sartoria, supported by Slam Jam
++ Nike
+++ Telecom Italia

Title of work
+ Deturno
++ F24
+++ Paidtelecom book and
packaging

Art Direction/Design
CD: Giorgio de Mitri
D: Gilles Cenazandotti

Type of work
+ +++ Media package
+++ Documentary book with
sculptural cardboard case

Specification
+ Use of material: Printed cardboard,
t-shirts, DVD, booklet
Dimension: 30 x 30 x 7 cm
++ Use of material: Plexi-glass tube,
cloth, cardboard, watercolor paper
and newsprint
+++ Use of material: Printed
cardboard
Dimension: 23 x 23 x 5 cm (assembled
30 cm high)

Origin
Modena, Italy

Year produced
+ 2001
++ 2004

Description
+ Defumo is the abbreviation of three artists' names: Delta, Futura and Mode2. The aim of the project was to bring three legendary graffiti artists under one roof and document both their process and the their dialogues about writing, colour, rhythm and art. The Defumo package included a documentary DVD, a limited edition silkscreen T-shirt by each artist and a booklet describing the project. The package was in the form of a hollowed out cube that opens in four sections.
++ FZ4 Media pack was conceived of as a tribute to Icons. The package, in the form of a plexi-glass tube, contained a large format journal with London based photographer Lawrence Passera's double exposure prints of Nike Total 90° icons and images created specifically for FZ4 by street icons: Delta, Futura and Mode2. In addition to the journal, the package contained a limited edition series of artists' prints, dedicated to iconic

Nike players Totti, van Neistelrooy, and Henry. The top of the plexi-glass tube doubled as the CD holder and the protective box in cardboard was lithographed by Futura.
+++ The Palatelecom book and cardboard package was created to document a year of Telecom Italia operations around Italy. For this occasion, Gilles Cenazandotti created a red cardboard projective case that became, by folding and inserting various cardboard pieces, the spitting image of the Palatelecom tower. The idea was to support the documentary of the Palatelcom events with a physical object that evoked the original architecture.

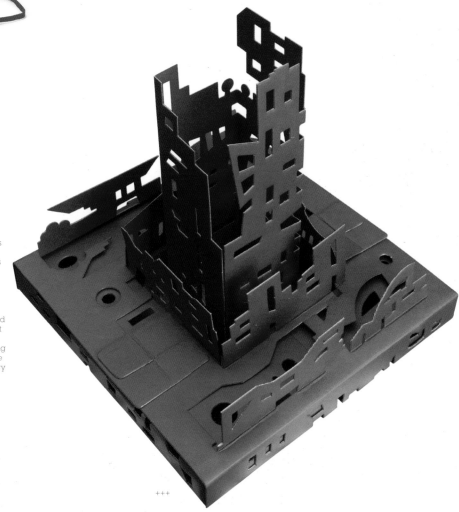

+++

+ Sweden Graphics

Page
152

Client
H&M

Origin
Stockholm, Sweden

Title of work
Divided Cosmetics Range

Art Direction/Design
Sweden Graphics

Year produced
2003

Type of work
Cosmetics Packaging

Specification
Use of material: Plastic
and metallic containers
Dimension: various

Description
The Divided Cosmetic range is
designed along the same principles
as the Divided Jeans and clothes
labels. The ambition is to create clear
and easily understandable product
groups, which also enable the manufacturer to add on further varieties
within every group as they become
available.

DIVIDED

03

wax
mousse

NET WT 4.62 OZ. /131 g

UNAVAILABLE

NET WEIGHT
1.7oz 50ml

™

+ Sagmeister Inc.

Page
154

Client
Blue Q

Origin
New York, USA

Title of work
Unavailable

Art Direction/Design
AD/D: Stefan Sagmeister
D: Hjalti Karlsson, Stephan Haas
CC: Karen Salmansohn

Year produced
2000

Type of work
Perfume packaging

Specification
Dimension: 4 3/4" x 6 3/4" x 1"

Description
Unavailable is a new fragrance
and soap line developed by Blue Q.
It is not only a perfume, it is also a
philosophy. The perfume is packaged
within a book that contains fifteen
Unavailable principles.

CONTENTS

THE INTRODUCTION
page 3

CLASSIC SUCCESS STORIES
A. ROMEO AND JULIET
B. CINDERELLA
page 5

2 PROVEN METHODS FOR APPLYING
UNAVAILABLE, THE PERFUME
page 7

15 PROVEN PRINCIPLES FOR APPLYING
UNAVAILABLE, THE PHILOSOPHY
page 9 – 23

THE FINE PRINT, THE FINER PRINT
& THE FINEST PRINT
page 25 – 27

Page
155

Title of work
Fragrance Range

Type of work
Packaging design

Client
Kangol

Art Direction/Design
AD: Harry Pearce
D: Harry Pearce, Christian Eves,
Rachael Dinnis

Specification
–

Origin
London, UK

Year produced
1999

Description
A range of fragrances designed for the
iconic fashion brand Kangol

KANGOL® TWO

BODY SPRAY

KANGOL fragrance for use all
over the body. Contains an
effective deodorant for long
lasting freshness.
Ingredients: Alcohol Denat, Butane,

EXTREMELY FLAMMABLE
Caution: Do not spray on a
naked flame or any incandescent
material. Keep away from sources
of ignition - No smoking.

200ml ℮

KANGOL

BODY SPRAY

+ Curiosity Inc.

Page
156

Title of work
+ Le Feu D'Issey
++ Equlir

Type of work
+ Perfume bottle
++ Essential oil bottle

Client
+ Beaute Prestige International
++ Kanebo Cosmetics

Art Direction/Design
+ AD: Issey Miyake
D: Gwenael Nicolas
++ AD, D: Gwenael Nicolas

Specification
+ Use of material: PCTA
Dimension: 70 x 70 x 90
++ Use of material: Glass,
and aluminium
Dimension: 38 x 38 x 60

Origin
Tokyo, Japan

Year produced
+ 1998
++ 2002

Description
This perfume bottle was designed to define the 'feel' of the bottle. To express the fire concept, a new material called PCTA was used to create a softer, warmer feel than the standard glass. This material was also appropriate for the complicated system inside and it expresses the softness of transparency and colour. The bottle is conceived for maximum ease. It is an automatic opening system, giving access to the spray by lifting the bottle, thus creating an intimacy feel through the interaction. ++ The cone shape is the ideal shape for a bottle and it certainly an air of elegance. The minute, dark-green crystals rock back and forth when being placed on a surface. With emphasis on the dynamic movement, the cap invites users to touch the bottle, while the three-layered package expresses the special and unique qualities of the product.

COMFORTABLE BODY VEIL   EQUILIR

宙

LISTEN TO YOUR NOSE   IT SENSES

™

+ Curiosity Inc.

Page
159

Title of work
Le Male Vapo Biscotto

Type of work
Packaging design

Client
Beaute Prestige International

Art Direction/Design
AD: Jean Paul Gaultier
D: Gwenael Nicolas

Specification
Use of material: Glass, plastic
Dimension: 35 x 70 x 90 mm

Origin
Tokyo, Japan

Year produced
2001

Description
This perfume is a figurative design of hand grippers used for grip strength training and the sports-oriented package features the shape of a dumb-bell. This approach works especially well with the imaginary and narrative messages used in accordance with the brand image of Jean Paul Gaultier. The innovative design is not only easy to use, but it also has a very natural scent.

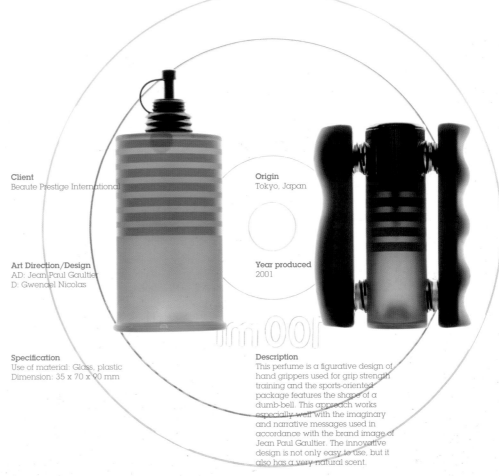

+ Curiosity Inc.

Origin
Tokyo, Japan

Title of work
Louis Vuitton shop invitation

Year produced
2003

Type of work
Invitation package

Specification
Use of material: Acrylic Resin
Dimension: 200 x 60 x 60

Description
This package was designed as the opening invitation for the main shop in Tokyo. The shop is situated in their new building and the acrylic object was a gift to their guests showing the image of the building. Following the concept of the building, the logo and marks were printed on the object to give a strange reflection, and the use of cube added an expectation for the guests to the shop.

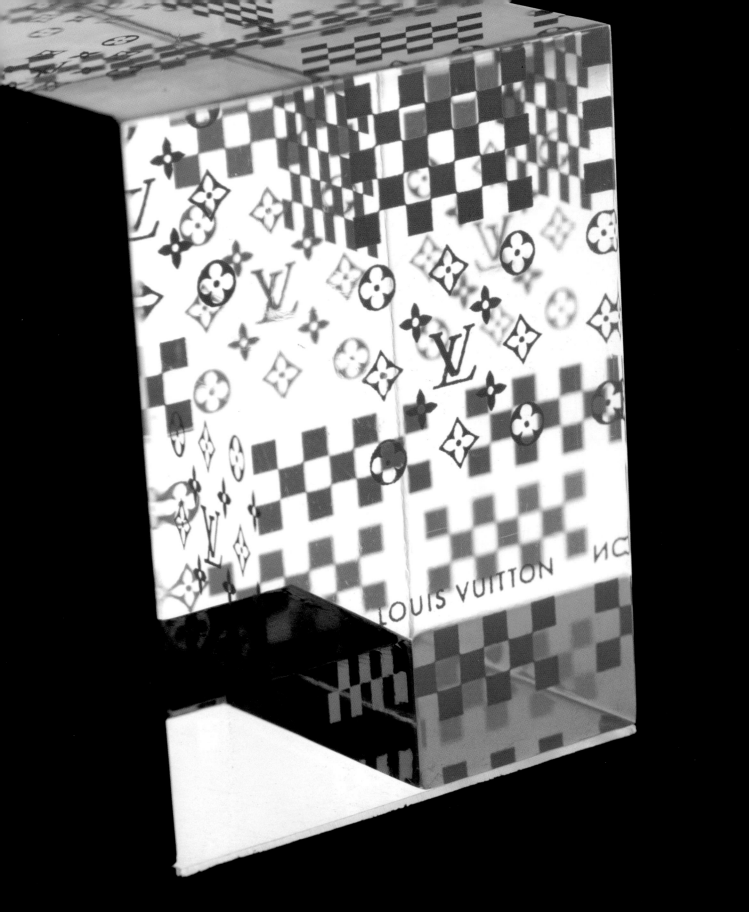

+ Curiosity Inc.

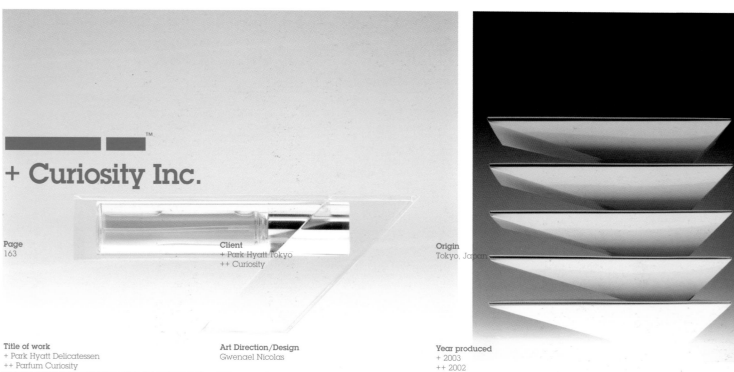

Client
+ Park Hyatt Tokyo
++ Curiosity

Origin
Tokyo, Japan

Title of work
+ Park Hyatt Delicatessen
++ Parfum Curiosity

Art Direction/Design
Gwenael Nicolas

Year produced
+ 2003
++ 2002

Type of work
+ Deli package
++ Packaging design

Specification
+ Use of material: Plastic and paper
Dimension: Various
++ Use of material: Acrylic glass
Dimension: 150 x 60 x 60 mm

Description
+ The white square motif shape is used on the shape for all items, highlighted with a ray of shadow, an obi "sleeve" in dark shadow green. The food is visible through the transparency inside the acryl boxes and protected by a paper wrapping.
The visual contrast of light and shadow is expressed in the tactile interaction, with sharp angle boxes placed into soft fabric bags.
The materials used are plain with minimum printing which reveals its taste and identity. A preview of the food quality gives a sense of authenticity and honesty.

++ The first view of Curiosity's perfume suggests the tension and character of the object. The purpose is to obtain a maximum of tension with a minimal of means since beauty is created by tension and silence. The object itself is a structural puzzle that maximizes the intrinsic beauty of the materials while the weight and structure create the overall balance. The scent container suspended in the centre is the structural heart of the object. The acrylic brings lightness, transparency and unexpected reflections. If viewed from a certain angle, the acrylic vanishes and only the liquid continues to be visible, thus realizing the designer's original inspiration.

++

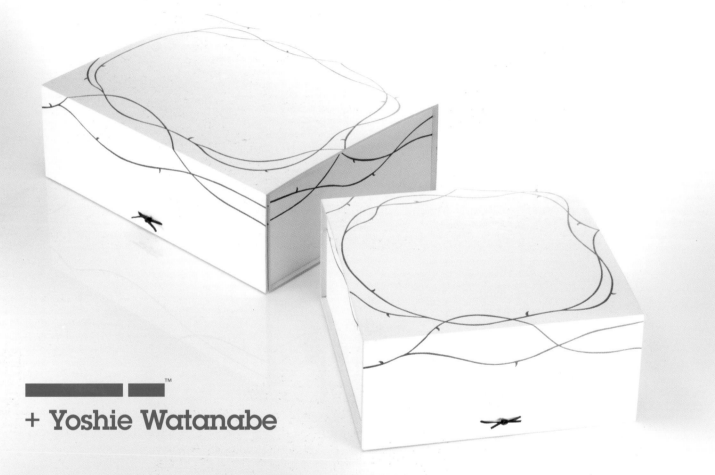

+ Yoshie Watanabe

Page
166

Client
Vitras Inc

Origin
Tokyo, Japan

Title of work
Mirror box

Art Direction/Design
Yoshie Watanabe

Year produced
2003

Type of work
Gift box

Specification
Dimension: Red – 205 x 205 x 95 mm;
Blue – 225 x 295 x 95 mm

Description
--

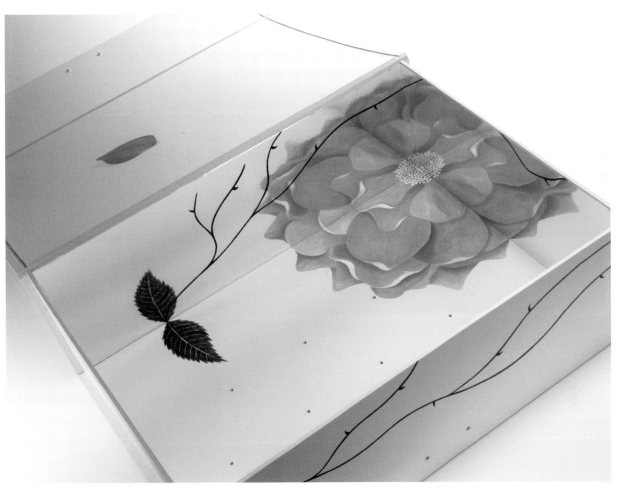

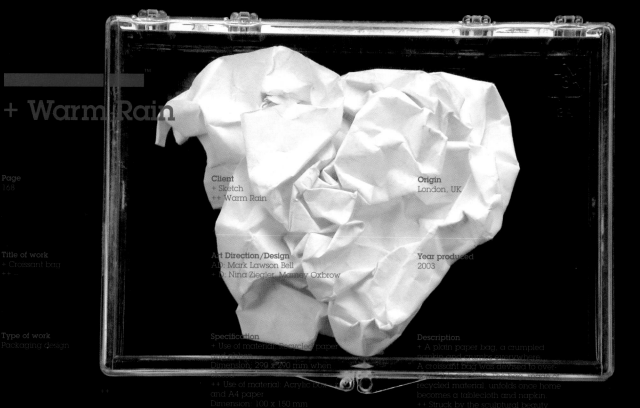

+ Warm Rain

Client
+ Sketch
++ Warm Rain

Origin
London, UK

Title of work
+ Croissant bag
++ ...

Art Direction/Design
AD: Mark Lawson Bell
+ D: Nina Ziegler, Marney Oxbrow

Year produced
2003

Type of work
Packaging design

Specification
+ Use of material: Recycled paper and ribbon
Dimension: 299 x 290 mm when open
++ Use of material: Acrylic box and A4 paper
Dimension: 100 x 150 mm

Description
+ A plain paper bag, a crumpled napkin and crumbs everywhere. A croissant bag was devised to over... the bag, made of recycled material, unfolds once home becomes a tablecloth and napkin
++ Struck by the sculptural beauty of a screwed up piece of paper and the fact that so many press releases end up in the bin. A press release was developed from this concept to save some people the job. Dazed and confused magazine described it as the best press release they had ever seen!

+ Yoshie Watanabe

Client
D-BROS

Origin
Tokyo, Japan

Title of work
Cardboard box

Art Direction/Design
Yoshie Watanabe

Year produced
1998

Type of work
Packaging design

Specification
Use of material: Paper

Description
–

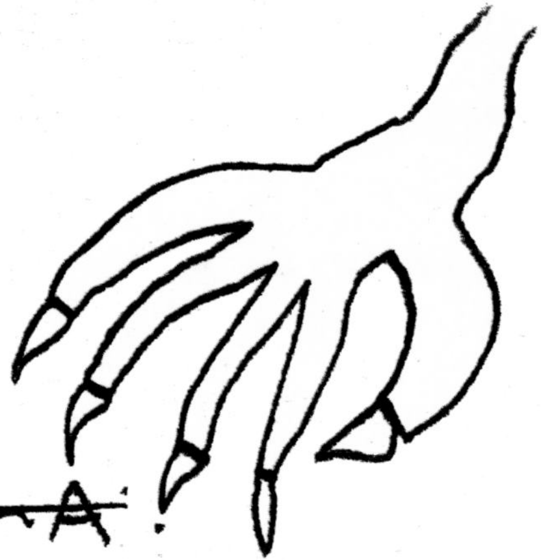

~~CLA~~!

CLOR

welcome music lovers

side a

1. good stuff

2. magic touch

side b

1. making you all mine

2. gifted

clor regal

www.clor.co.uk

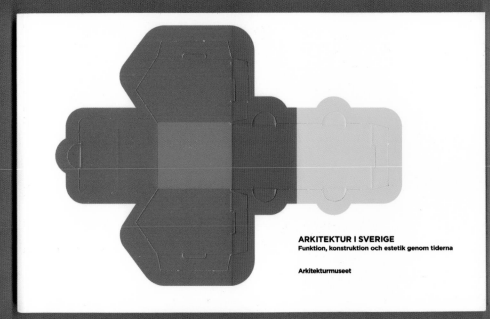

ARKITEKTUR I SVERIGE
Funktion, konstruktion och estetik genom tiderna

Arkitekturmuseet

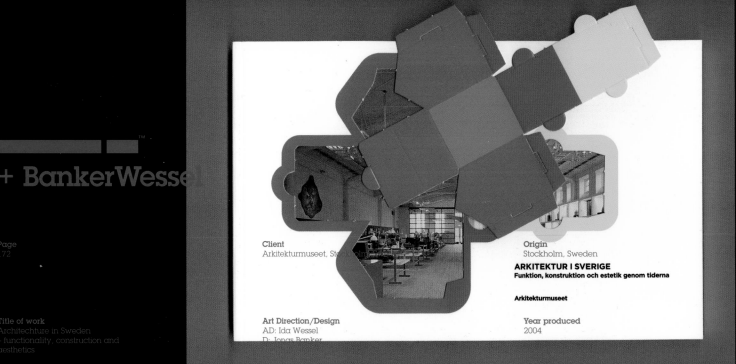

Client
Arkitekturmuseet, Stockholm

Origin
Stockholm, Sweden

ARKITEKTUR I SVERIGE
Funktion, konstruktion och estetik genom tiderna

Arkitekturmuseet

Art Direction/Design
AD: Ida Wessel
D: Jonas Banker

Year produced
2004

Title of work
Architechture in Sweden
– functionality, construction and
aesthetics

Type of work

Specification

Description

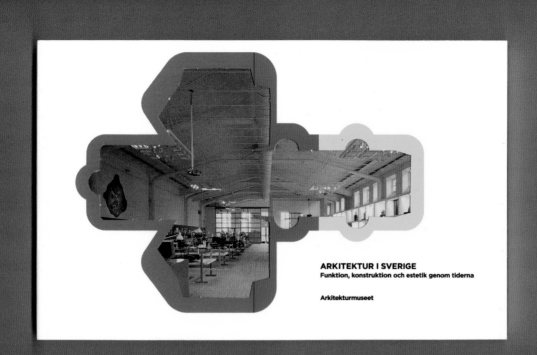

ARKITEKTUR I SVERIGE
Funktion, konstruktion och estetik genom tiderna

Arkitekturmuseet

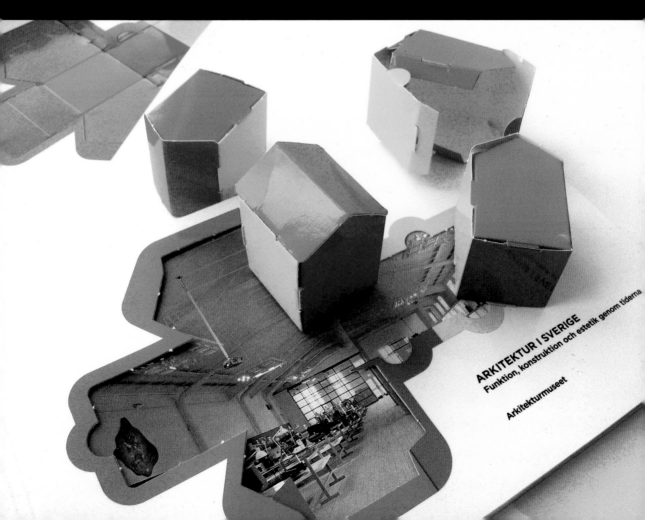

™

+ Segura Inc.

Page	**Client**	**Origin**
174	--	Chicago, USA
Title of work	**Art Direction/Design**	**Year produced**
Express Jeans – Dangerously Bold	Carlos Segura	2004
Type of work	**Specification**	**Description**
Tags	--	+ The graphic is silk-screened on felt or wool material. A transparent piece of coloured vinyl material is stitched to the back with one end left open to form a pocket. The car image is printed on the back of the fabric which is shown through the vinyl. Information is silk-screened on top of the vinyl.

+ The graphic is silk-screened on felt or wool material. A transparent piece of coloured vinyl material is stitched to the back with one end left open to form a pocket. The car image is printed on the back of the fabric which is shown through the vinyl. Information is silk-screened on top of the vinyl.
++ The graphics are silk-screened and stitched on the cloth tag with a button. Once removed, it can then be used as a carrier for a PDA, I-POD or mobile phone.
+++ The template has a combination of cut-out icons with graphics printed in black on a standard light green plastic template with a ball and chain fastener. The customer can retain the tag and use it in a variety of ways to make their marks.

TWO PIECES STICHED TOGETER TO FORM POCKET

+ Hans Seeger

Page
176

Client
Aesthetics

Origin
USA

Title of work
Pulseprogramming – Tulsa for one
second:
+ CD/LP
++ Remix project no.1 – Schneider TM
& Static
+++ Remix project no.2 – Hood,
Barbara Morgenstern and Köhn

Type of work
Music packaging

Art Direction/Design
AD: Hans Seeger, John Shachter
+ Pt: Blanchette Press
++/+++ P: Ola Bergengren

Specification
+ Use of material: Offset on tyuek
Dimension: Various
++/+++ Use of material: Offset on
cardstock
Dimension: 12.125″ x 12.125″

Year produced
+ 2003
++/+++ 2004

Description
+ A full-length release from Chi-
cago-based Pulseprogramming. The
packaging reflects the record's traits
of warmth, hospitality and humanity
through the concept of home.
++ The limited edition, vinyl-only
remix series based on Pulseprogramming's full-length release "Tulsa for
one second". The concept was to have
the contributing artists remix both the
audio track and the original house
packaging. The results were then pho-
tographed by Ola Bergengren. This
first release also contained the work
of Schneider TM and Static.

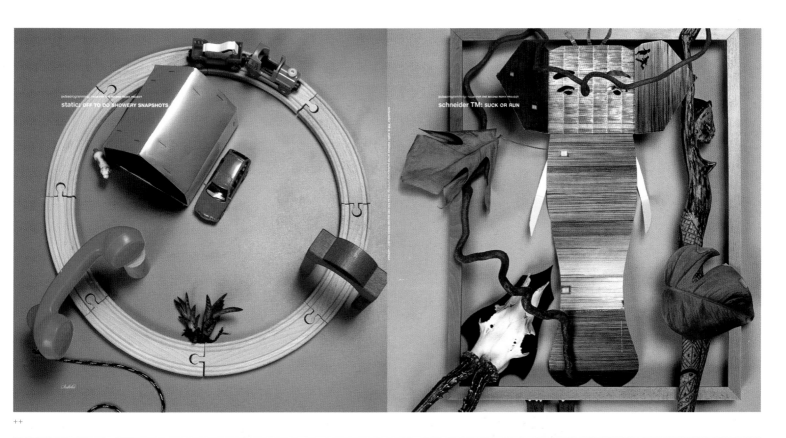

pulseprogramming: TULSA FOR ONE SECOND REMIX PROJECT
static: OFF TO DO SHOWERY SNAPSHOTS

pulseprogramming: TULSA FOR ONE SECOND REMIX PROJECT
schneider TM: SUCK OR RUN

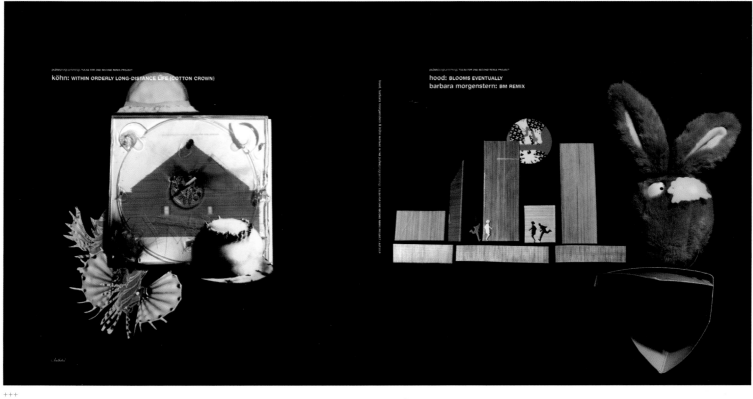

pulseprogramming: TULSA FOR ONE SECOND REMIX PROJECT
köhn: WITHIN ORDERLY LONG-DISTANCE LIFE (COTTON CROWN)

pulseprogramming: TULSA FOR ONE SECOND REMIX PROJECT
hood: BLOOMS EVENTUALLY
barbara morgenstern: BM REMIX

++

+++

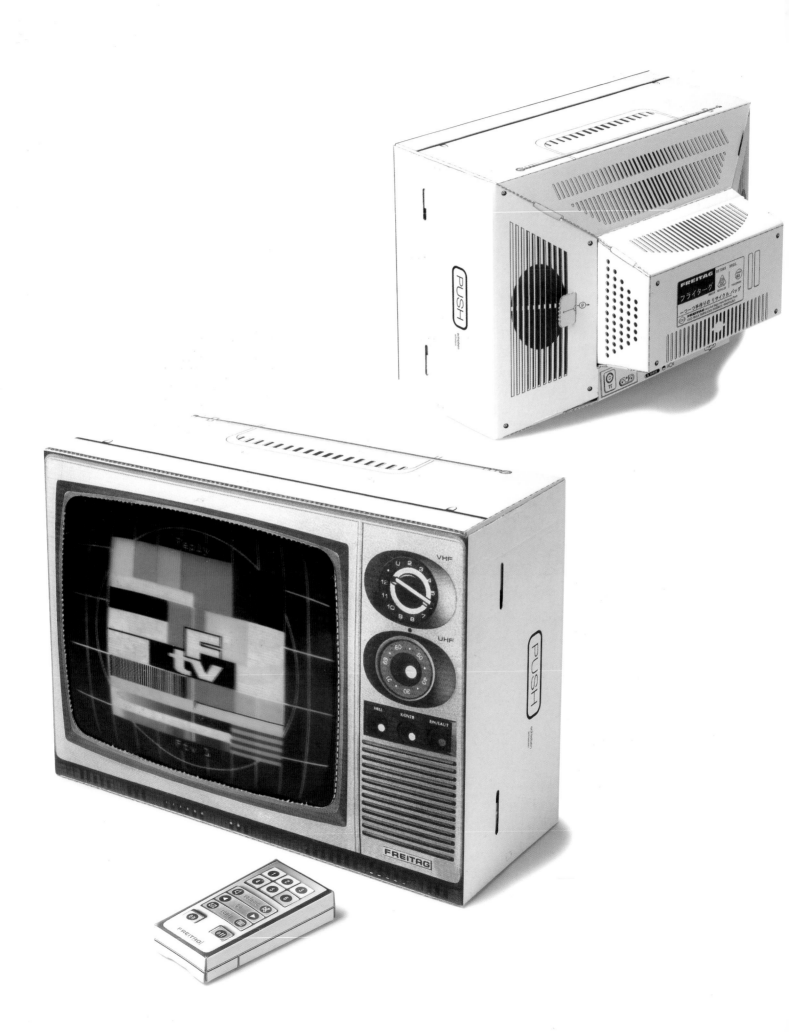

+ Freitag ™

Page
179

Client
FREITAG

Origin
Zürich, Switzerland

Title of work
FTV box

Art Direction/Design
AD: Daniel Freitag, Markus Freitag
D: Daniel Freitag

Year produced
1994

Type of work
Packaging for FREITAG bags

Specification
Use of material: Cardboard
Dimension: 248 x 435 x 95

Description
This is the FREITAG classic box - one box fits almost all. Since no FREITAG bag is like another, there is a photo of each individual bag on the outside of the box. Stacked the F-way, the boxes turn into a presentation wall, where the customer can see the interesting designs or photos and pick up the real bag by opening the box like a drawer. Every box can be assembled into a decorative fake TV, and after sending in the response card, FREITAG would deliver the F-TV remote control for free.

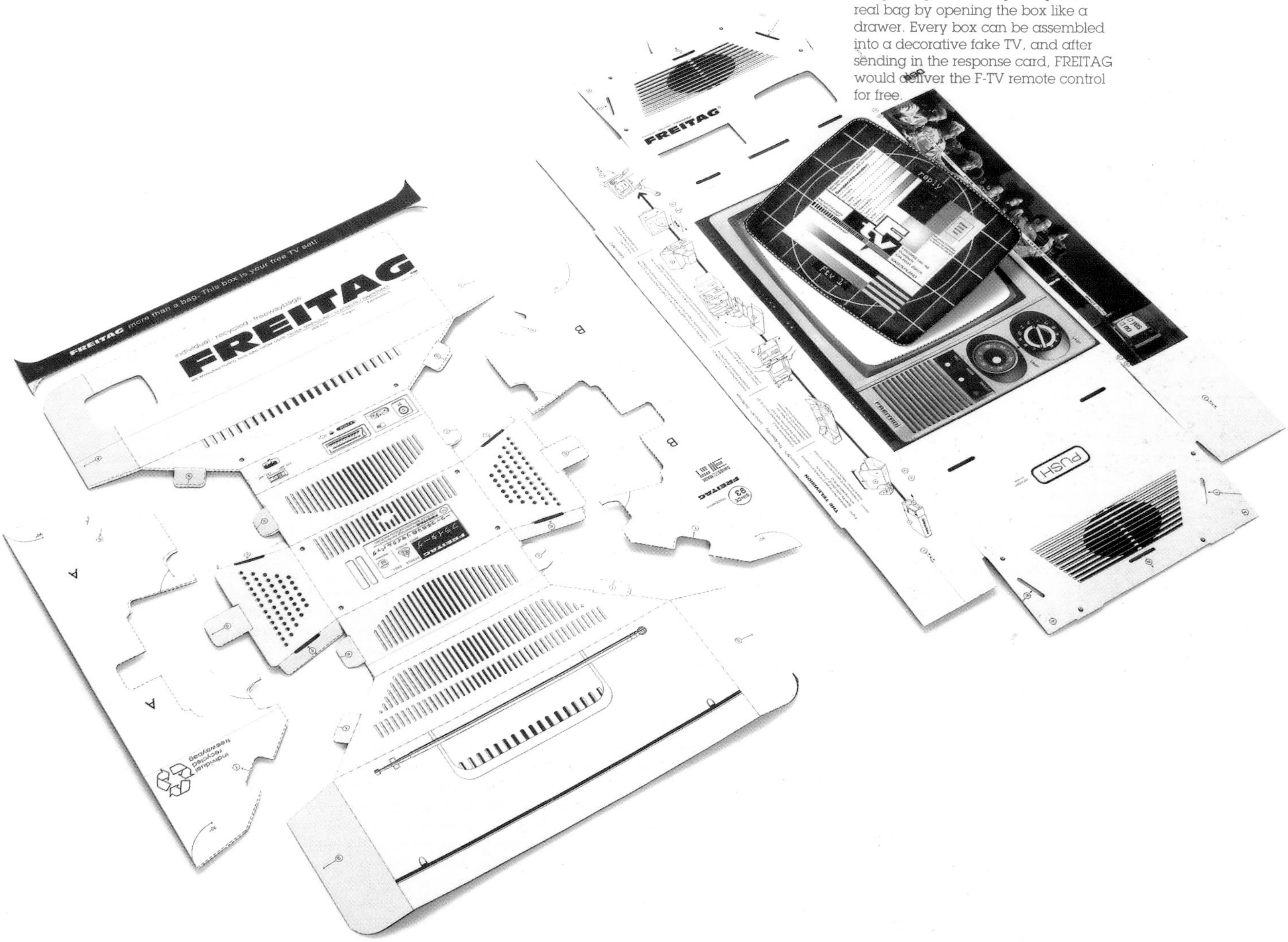

+ Freitag ™

Title of work
+ Twinpeaks
++ F-CASH wallet displays
+++ F69-Calcio

Type of work
+ Packaging for FREITAG bags F71-74
++ Packaging for FREITAG wallets
F50, F51+F53
+++ Packaging for FREITAG footballs

Art Direction, Design
+ AD: Daniel Freitag
Markus Freitag
Ralph Steinbrüchel
++ AD: Daniel Freitag
Markus Freitag
D: Luca Rastello

Specification
Use of material: Cardboard
+ Dimension: 765 x 397.5 x 464 mm
++ Dimension: 240 x 436 x 44 mm
+++ Dimension: 218 x 218 x 222 mm

Origin
Zürich, Switzerland

Year produced
+ 2003
++ 2004
+++ 2001

Description
+ The box is used to ship one dozen
FREITAG Twinpeaks bags to retailers.
The outside of the box features slot-ted
cut-lines for ready-to-use bag-storage.
This ensures a perfect display of
FREITAG's LAURA, DONNA, COOPER
and BOB bags in the stores.
++ To ensure a more compact display
of FREITAG wallets in stores, it was
decided that they would be delivered
in a ready-made display box. Up to
eighteen units of FREITAG E.T., DALLAS
and DENVER models can be stored in a
relatively small place, giving customers
the opportunity to choose between the
individual, mouth-watering colours and
designs.
+++ Round footballs are tough to
stock, so this packaging makes them
square. In addition to a good view on
the individual design of the ball, the
FREITAG CALCIO box comes with an
array of cut-out cardboard football
accessories: peanuts, a whistle, a
baseball bat, band aids, a red and a
yellow card, as well as the obligatory
bottle of beer (the only kind allowed in
the stadium...).

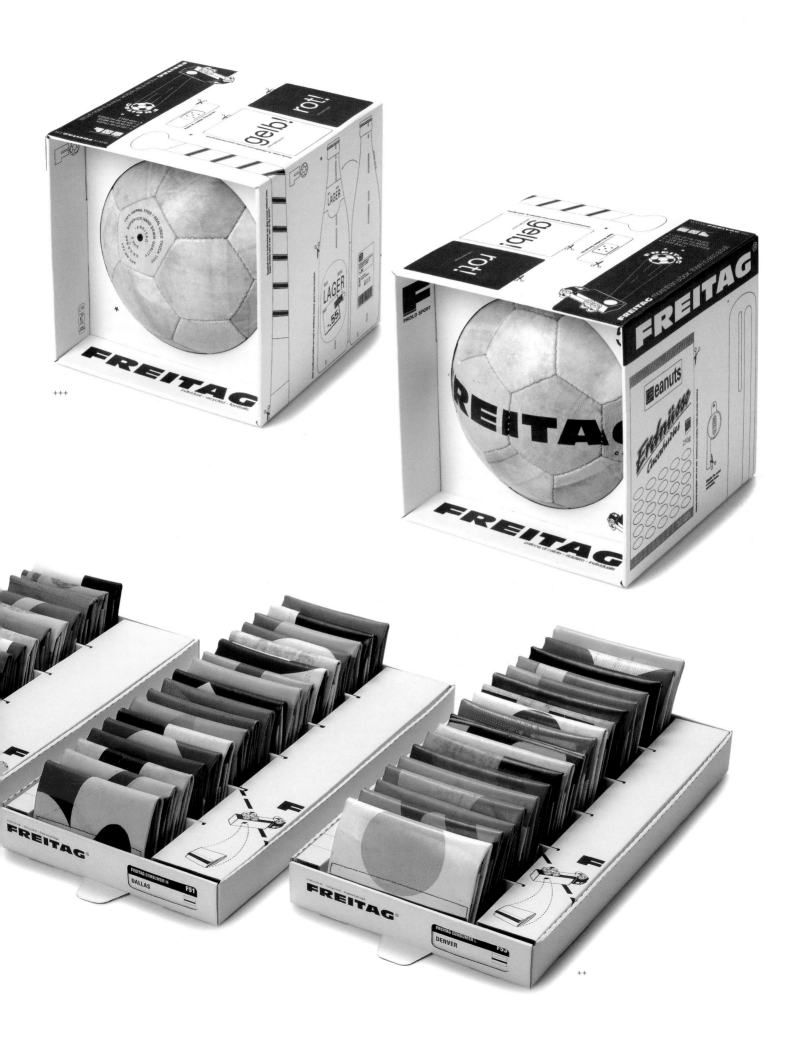

+ Satellite Design ™

Page
182

Client
The North Face

Origin
California, USA

Title of work
The North Face Tent and Sleeping Bag
Packaging

Art Direction/Design
Amy Gustincic

Year produced
2003

Type of work
Retail camping equipment packaging

Specification
Use of material: Collection-specific
corrugated boxes with litho-lam
printing and product-specific offset-
printed
Dime   35.5 x
30.5   x 30 x
24.

Description
Responding to requests from dealers,
The North Face began packaging
its tents and sleeping bags. The
package design posed problems of
cost, merchandising and shipping
efficiencies. This solution uses
collection-specific boxes and product-
specific labels to control costs without
sacrificing product information. A
stackable hexagonal shape and
strong brand images help the
packaging stands out at retail stores.

™

+ Bleed

Page
183

Client
+ Levi´s Nordic
++ djuice

Title of work
+ Icon Fightnight
++ djuice giveaway box

Art Direction/Design
+ Dag Solhaug
++ AD/D: Erik Hedberg,
AD: Svein H. Lia

Year produced
+ 2003
++ 2004

Type of work
+ Packaging/invitation
++ This is a promotion box produced
for the telecom company djuice. The
box was given to visitors at the Quart
Festival, the largest music festival in
Norway, which was partly sponsored
by djuice. The box contained a dispos-
able grill, a package of hot dogs, a
t-shirt, and some other goodies like
aspirin and condoms...

Specification
+ Use of material: Basic milk packag-
ing
Dimension: 0.5 Litres, 10 x 10 x 17 cm
++ Use of material: Cardboard
Dimension: 275 x 345 x 123 mm

Description
+ Invitations made for Levi´s tron online
Release Training Session Spring 2003
were sent out to key account customers. Focus for the evening was Icon
Fightnight (IF), where the guests had to
do verbal "fights" on knowledge about
the Icon´s (Levi´s) new collection. The
invitations were sent out in milk cartons.
Inside, you would get different kind of
items, such as a pair of boxing gloves
attached to a key ring.
++ The graphics on the box are based
on the "look" created for djuice, who
sponsored several festivals in Norway
during the summer of 2004. They sym-
bolize music, happiness, freedom and
summer.

++

+ unfolded ™

Page
184

Title of work
Tommy Fjordside Catalogue Winter
04/05

Type of work
Catalogue

Client
Tommy Fjordside

Art Direction/Design
CD: Nadia Gisler
AD: Friedrich-Wilhelm Graf
P: Nicole Bachmann

Specification
Use of material: Paper - Munken Lynx
Print - Four CMYK + One pantone
Dimension: Folded - 12.5 x 12.5 cm
Unfolded - 50 x 50 cm

Origin
Zürich, Switzerland

Year produced

Description
For the winter 04/05 catalogue of
the Tommy Fjordside fashion design
studio, we decided to
a fold-out
the tight bu
Winter was a
the special fold (flip
and pictures) so the catalogue
was hand-folded
specializes in the
handicrafted pa
The folding me the tactile
of Tommy Fjordside's cloth

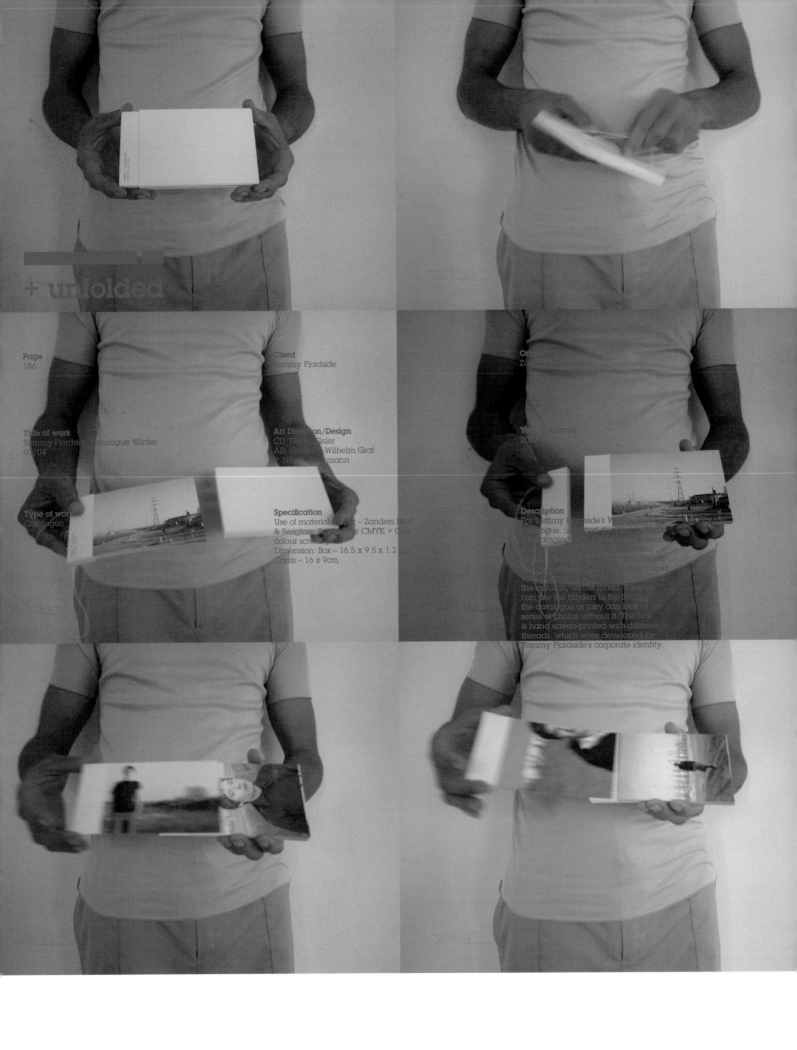

Page
186

Client
Tommy Fjordside

Origin
Zürich

Title of work
Tommy Fjordside Catalogue Winter
03/04

Art Direction/Design
CD: Nadine Gisler
AD: Friedrich-Wilhelm Graf
P: Nicola Schumann

Year
2003

Type of work
Catalogue

Specification
Use of material: paper – Zanders Matt
& Serigloss; Print colour CMYK + One
colour screen-print.
Dimension: Box – 16.5 x 9.5 x 1.2 cm
Cards – 16 x 9cm

Description
For Tommy Fjordside's Winter
catalogue, unfolded designed the
packaging. The catalogue is a series
of photographs, which show cinematic
and romantic images, which are in
keeping with the music. They reflect
the autumn/winter mood. Viewers
can use the binders to flip through
the catalogue or they can look at the
series of photos without it. The box
is hand screen-printed with different
threads, which were developed for
Tommy Fjordside's corporate identity.

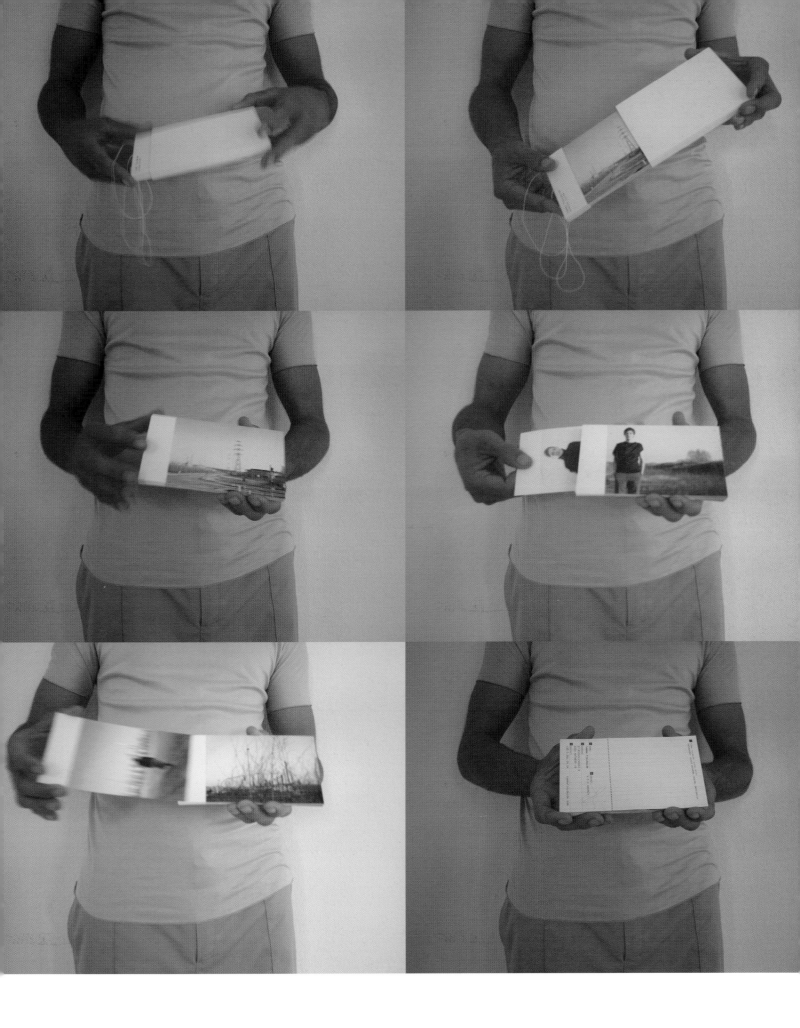

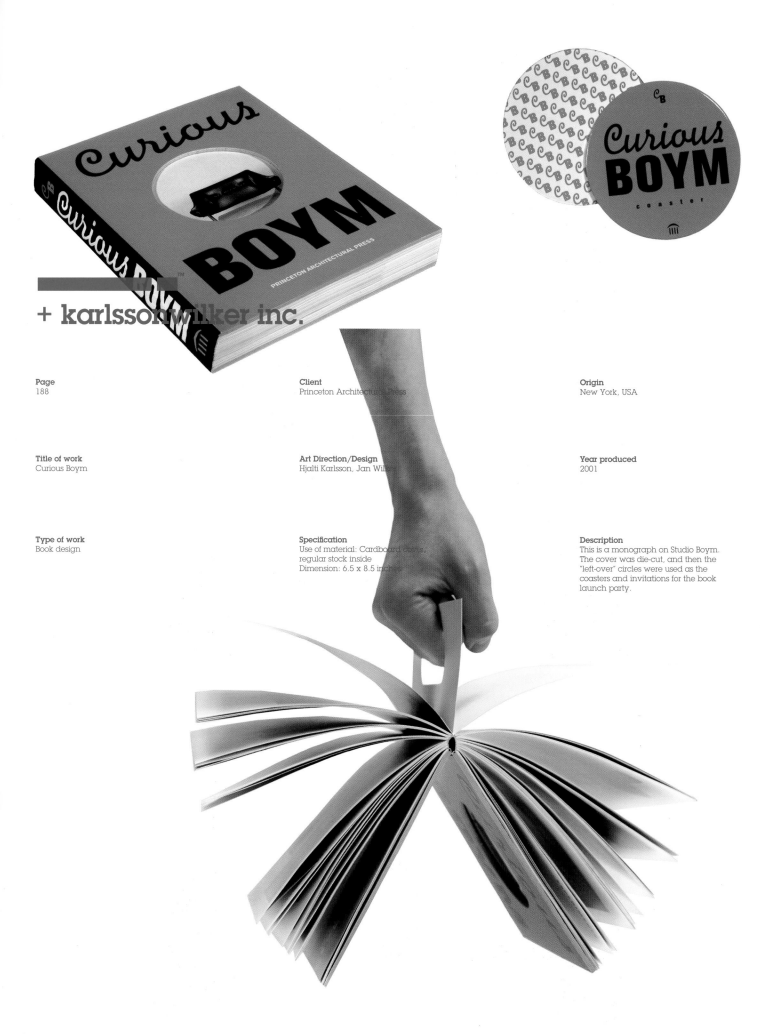

+ karlssonwilker inc.

Page
188

Client
Princeton Architectural Press

Origin
New York, USA

Title of work
Curious Boym

Art Direction/Design
Hjalti Karlsson, Jan Wilker

Year produced
2001

Type of work
Book design

Specification
Use of material: Cardboard cover,
regular stock inside
Dimension: 6.5 x 8.5 inches

Description
This is a monograph on Studio Boym.
The cover was die-cut, and then the
"left-over" circles were used as the
coasters and invitations for the book
launch party.

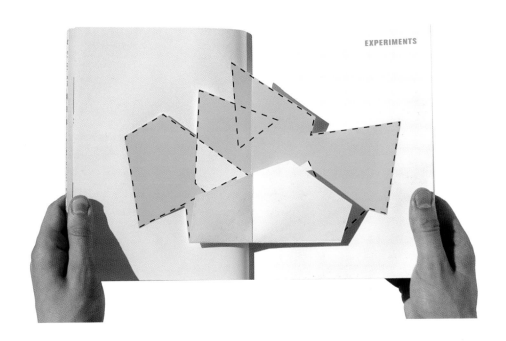

EXPERIMENTS

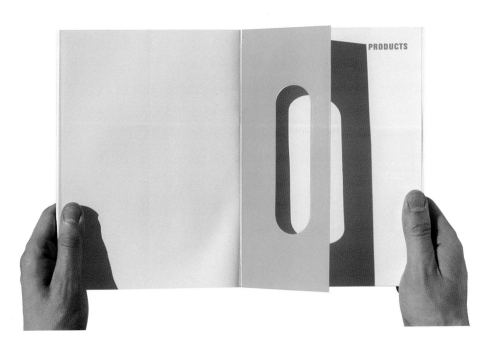

PRODUCTS

ENVIRONMENTS

+ strichpunkt

Page
190

Client
4MBO AG, Plochingen

Origin
Stuttgart, Germany

Title of work
Shopping bag

Art Direction/Design
CD: Kirsten Dietz, Jochen Rädeker
D: Stephanie Zehender

Year produced
2003

Type of work
Annual Report 2002

Specification
Use of material: Shopping bag
Dimension: 21 x 26 cm

Description
4MBO sells high-tech products in large amounts through low-cost channels, mainly supermarkets. The annual report features the 4MBO clients by using their logos on the shopping bags. In order to emphasize the successful partnership, the names of the clients are "completed" in business terms: e.g. "extra" to "real" to "sense of realism", "spar" (German for save) to economy drive (spareffekt).

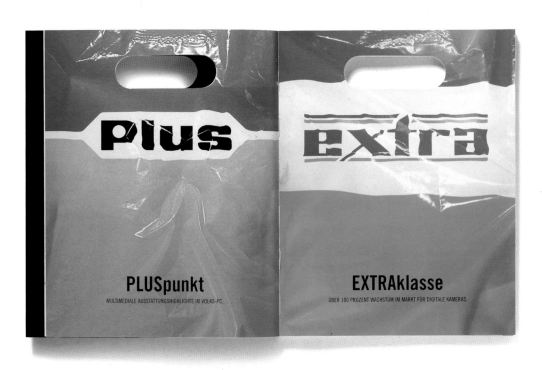

INHALT

VORWORT DES VORSTANDES // S 13
DIE 4MBO-AKTIE // S 17
KONZERN-LAGEBERICHT // S 21
AG-LAGEBERICHT // S 47
BERICHT DES AUFSICHTSRATES // S 57

+ strichpunkt

Client
CANCOM AG

Origin
Stuttgart, Germany

Title of work
—

Art Direction/Design
CD/D: Kirsten Dietz
CD: Jochen Rädeker

Year produced
2000

Type of work
Annual Report 1999

Specification
Use of material: Corrugated cardboard
and antistatic shield cover

Description
CANCOM is a provider of individually
configured PC hard and software for
creative businesses like advertising
agencies and designers. An annual
report brochure helps to build an individual image of the company featuring its product advantages.
Packed in corrugated cardboard and
antistatic shield cover, the brochure
is packed and delivered exactly the
same way as the CANCOM products
and is assembled in the same packing
compartment of the company.

...NNAISE ___ SONDERN MIT BERATUNG

CANCOM SETZT NICHT AUF DIE RIESEN-HAUSHALTSPACKUNG ___ ZUM JUBILÄUMSPREIS ___ SONDERN AUF SERVICE

Starke Kennzahlen

Die von 47,4 Mio. DM auf 130,7 Mio. DM deutlich verlängerte Bilanzsumme zeigt eindrucksvoll das enorme Wachstum des CANCOM-Konzerns. Unsere beachtliche Eigenkapitalquote des Vorjahres von 47,8 Prozent erhöhte sich nochmals und lag zum Bilanzstichtag bei 49,8 Prozent. Eine Anlagendeckung von nahezu 181 Prozent und Bankverbindlichkeiten von gut 0,7 Prozent des Umsatzes belegen unsere ausgezeichneten horizontalen und vertikalen Bilanzrelationen und unsere hohe Liquidität.

SOFTMAIL GIBT ES WEDER GEZUCKERT NOCH MIT ___ SCHOKOLADE UMHÜLLT ___ SONDERN PER MAIL

+ strichpunkt

Ignition

Page
194

Client
Daimler Chrysler AG, Stuttgart

Title of work
Ignition / thinking in ideas

Art Direction/Design
AD: Jochen Rädeker, Kirsten Dietz
D: Felix Widmaier

Type of work
Brochure / matchbox

Specification
Use of material: Various
Dimension: 28.5 x 35 cm

AKTIVES
KURVENLICHT

Origin
Stuttgart, Germany

Year produced
2003

Description

A brochure created for the IAA 2003 presented the new technical features of the current Mercedes-Benz automobile range. It used striking images to create an emotionally charged stimulus.

As a matter of principle, brilliant ideas are simple, however high-tech solutions for safety, driving enjoyment and comfort are highly complex. Their effect and operation on the other hand are immediately clear. As clear as striking a match, being the driver of a top model car is easy to explain. Seven striking ideas that fire the imagination are authentically put across in seven premium locations. Matching the fold-out photos of the new Mercedes models in design and colour and including additional pages with details. As striking as the power released by a match or an ignition key. Produced in the same high quality as a Mercedes: in hardcover, with special colours and special paints, and compactly packed in a matchbox.

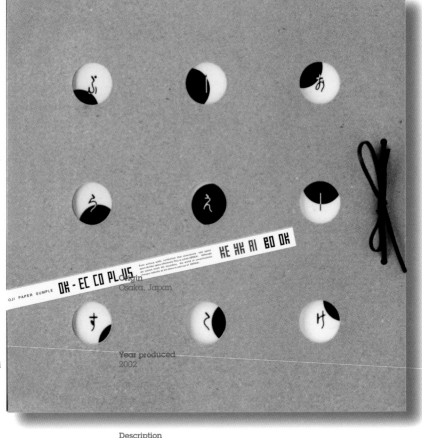

▬▬▬▬▪™
+ Kokokumaru

Page
196

Client
Oji paper Co. Ltd

Origin
Osaka, Japan

Title of work
KEKKAI Book

Art Direction/Design
Yoshimaru Takahashi

Year produced
2002

Type of work
Book packaging

Specification
Use of material: Paper sample book
Dimension: 300 x 300 x 10 mm

Description
--

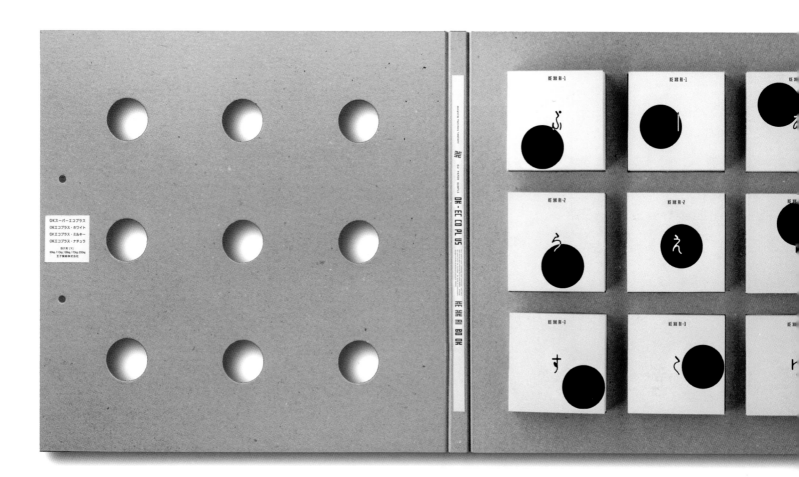

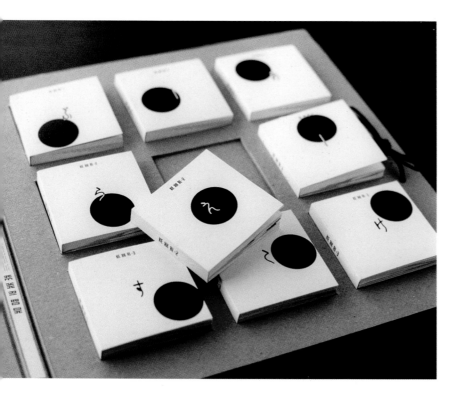

Page
198

Client
IdN

Origin
Hong Kong

Title of work
My Favourite Conference 2004 –
"Hook On Your Favourite"

Art Direction/Design
Victor Cheung

Year produced
2004

Type of work
Conference value-pack

Specification
Use of material: Paper
Dimension: 14.5 x 18.5 x 1.7 cm

Description
This clean approach for the value-pack contains a booklet, CD-Rom and a conference entry pass. The unique way of binding is used in order to seal all items in one pack as security purposes for retailers.

+ Époxy

Page
200

Client
Marie-St Pierre (fashion designer)

Origin
Montreal, Canada

Title of work
Marie-St Pierre Tube

Art Direction/Design
CD : Daniel Fortin, George Fok
AD: Marie-Hélène Trottier
D: Marie-Hélène Trottier
PdM: Hélène Joanette
PjM: Marie-geneviève Parent
CG: André Renaud

Type of work
Packaging design

Specification
—

Description

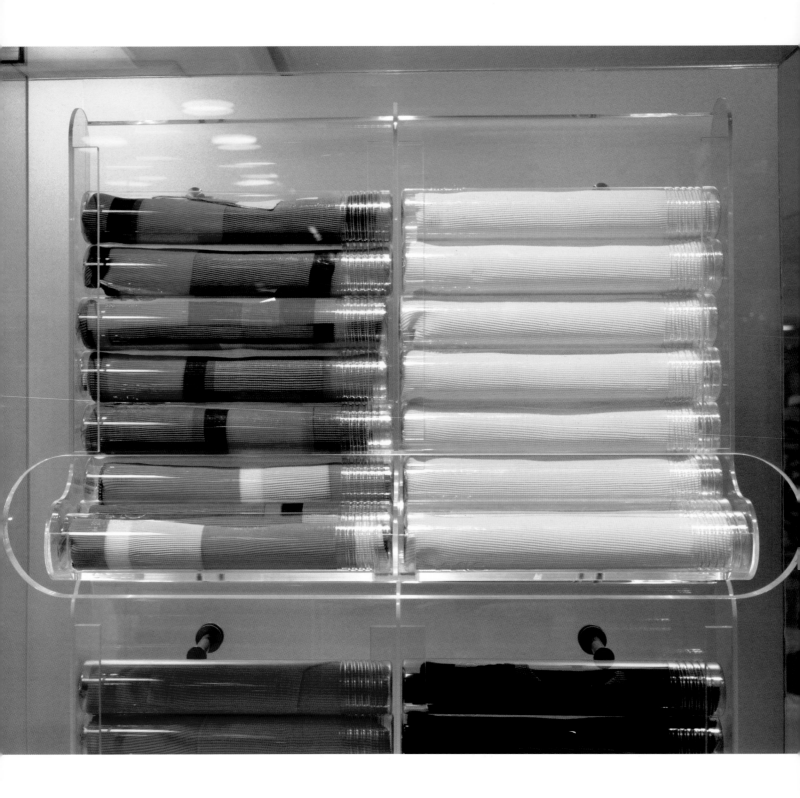

+ Curiosity Inc.

Page
203

Client
Issey Miyake

Origin
Tokyo. Japan

Title of work
Me Issey Miyake

Art Direction/Design
Gwenael Nicolas

Year produced
2001

Type of work
Clothing package

Specification
Use of material: PET
Dimension: 40 x 40 x 370

Description
"Me" clothing are exclusively tops. They have the ability to stretch widely so one size can fit everybody. This unique sizing provided an opportunity to develop a new concept for the shop and the package. The clothing is packaged from the factory in a transparent plastic tube and is displayed in a transparent machine. The tubes were designed specifically to work with this display. The plastic (PET) tube itself, with a large screw to open on one side, is like a mineral water bottle, transparent with the logo embossed on the front. The tube can be taken from the front of the display, allowing the next tube to appear automatically.

+ Doyle Partners

Page
204

Client
Gael Towey / Martha Stewart Living
Omnimedia; Kmart
New York, USA

Title of work
Martha Stewart Everyday Packaging
for Kmart

Art Direction/Design
CD: Stephen Doyle
D: Tom Kluepfel, Rosemarie Turk, Lisa
Yee, Ariel Apte, Vivian Ghazarian,
John Clifford, Liz Ahrens, Gratia Gast,
Michelle Cosentino, Jia Hwang, Craig
Clark, Naomi Mizusaki, Elizabeth Lee,
Vanessa Eckstein
PM: Cameron Manning, Goizalde
Mintegia

Year produced
1996 - 2000

Type of work
Packaging

Specification

Description
As a brand, Martha Stewart Everyday is equal parts information and inspiration; the package design brings these core concepts to the retail experience. Simple typography, striking colours and easy, accessible design show each product to its best advantage and enliven the shopping experience. Packaging was constructed to highlight the product attributes, allowing light to shine through the glassware packages, letting customers feel the weight of flatware. The pre-sentation of uninterrupted surfaces of plates in a subtle range of colours also conceived to give customers a better understanding of the products' design properties. In a fluorescent, mass shopping environment, the authority and exuberance of these packages commands consumer attention and trust — and record breaking sales.

+ strichpunkt

Page
206

Client
strichpunkt

Origin
Stuttgart, Germany

Title of work
Festessen

Art Direction/Design
CD/D: Kirsten Dietz
D: Gernot Walter

Year produced
--

Type of work
A Christmas feature on roast goose

Specification
--

Description
A Christmas mailing on the subject of roast goose replaces the classic image brochure. Instead of delivering samples of work and noble words, the customers are called to action since interaction is the basis of successful cooperation. In addition to conveying wit and the fun aspect, the typical strichpunkt typography and illustration are also shown.

+ strichpunkt™

Title of work
Trading on the Net

Type of work
Annual Report

Client
CANCOM AG

Art Direction/Design
CD/D: Kirsten Dietz
CD: Jochen Rädeker

Specification
Use of material: Net

Description
CANCOM is the market leader for DTP products (Apple/...) and their products are sold... their newly developed online shops. Their target markets are shareholders (Neuer Markt investors) and customers (in particular advertising agencies). The "Trading on the Net" theme communicates the advantages of CANCOM shops over classic retailing using a juxtaposition of images and text (e.g. Service – Apple in a supermarket, Consulting – software at a fast food outlet, low price – Hardware at the chemist). Supermarket elements are used in texts and "diagrams", delivering very sound and serious information.

Year produced
2001

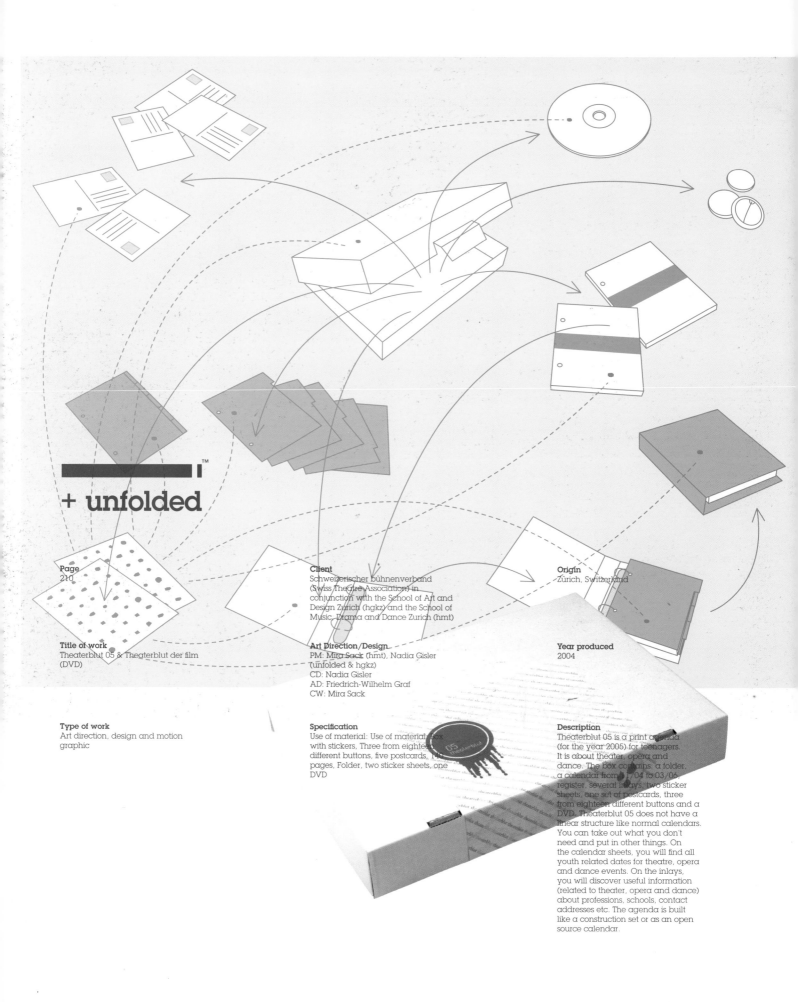

+ unfolded

Page
210

Title of work
Theaterblut 05 & Theaterblut der film
(DVD)

Type of work
Art direction, design and motion
graphic

Client
Schweizerischer bühnenverband
(Swiss Theatre Association) in
conjunction with the School of Art and
Design Zurich (hgkz) and the School of
Music, Drama and Dance Zurich (hmt)

Art Direction/Design
PM: Mira Sack (hmt), Nadia Gisler
(unfolded & hgkz)
CD: Nadia Gisler
AD: Friedrich-Wilhelm Graf
CW: Mira Sack

Specification
Use of material: Use of material box
with stickers, Three from eighteen
different buttons, five postcards, MG
pages, Folder, two sticker sheets, one
DVD

Origin
Zürich, Switzerland

Year produced
2004

Description
Theaterblut 05 is a print agenda
(for the year 2005) for teenagers.
It is about theater, opera and
dance. The box contains: a folder,
a calendar from 01/04 to 03/06,
register, several inlays, two sticker
sheets, one set of postcards, three
from eighteen different buttons and a
DVD. Theaterblut 05 does not have a
linear structure like normal calendars.
You can take out what you don't
need and put in other things. On
the calendar sheets, you will find all
youth related dates for theatre, opera
and dance events. On the inlays,
you will discover useful information
(related to theater, opera and dance)
about professions, schools, contact
addresses etc. The agenda is built
like a construction set or as an open
source calendar.

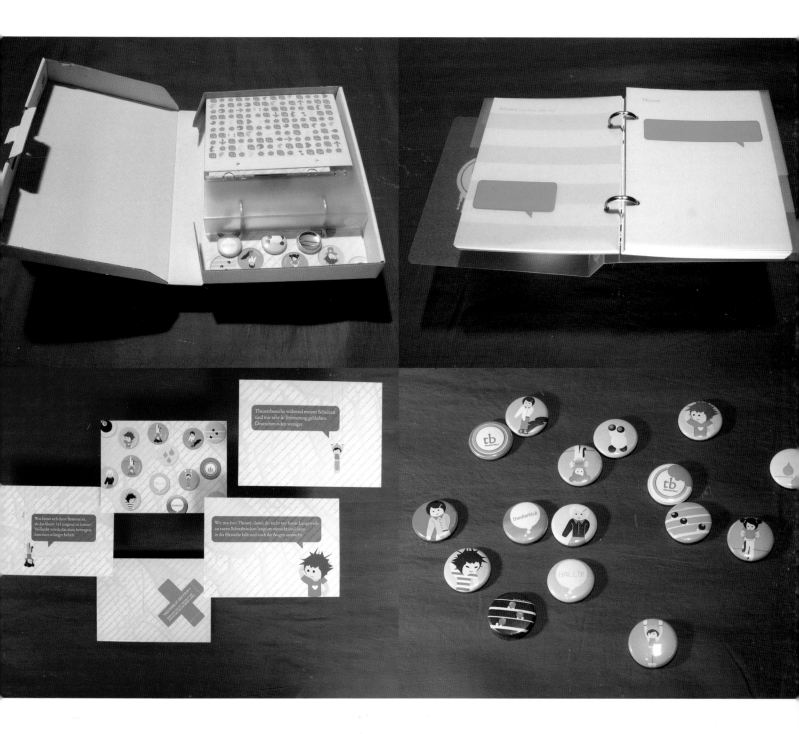

+ BrandIgloo™

Page
212

Title of work
SecretDiary

Type of work
A website diary sold as a real packaging in stores

Origin
Hamburg, Germany

Year produced
2003

Description
The Secret Diary™ is a private online diary for two people, it allows you to overcome any physical distance to your nearest and dearest and express what you feel for each other. Whenever there is something you would like to tell your Secret Diary™ partner, whether it is something special or you are just thinking of them, you can share this experience in an unique way. Like a real diary, it allows you to reveal your most intimate thoughts to each other – over time it will become an everlasting personal document of your time together, which can be obtained as a bound hardcover at any time.

+ Crush ™

Client
Crush

Origin
Brighton, UK

Title of work
Crush Box

Art Direction/Design
AD: Carl Rush
D: Simon Slater

Year produced
2002

Type of work
Self-promotional work

Specification
Use of material: Wooden box made
from found wood and screen-printed
Dimension: 12" square/3" deep

Description
The box contained two screen-printed
T-shirts and postcard pack. When
postcards are put together, they create a poster. The box is nailed shut,
and can only be opened with a crow-
bar. Sticker around the box is used
as an address label. The boxes were
made in a limited edition run of thirty
and were sent to clients and potential

+ Crush ™

Client
+ Full On Film Productions
++ Palm Pictures
+++/++++ Urban Theory

Origin
Brighton, UK

Title of work
+ Your Gonna Wake Up One Morning
++ Quantum Dub Force album packaging
+++ DJ Spinna
++++ Sex Sluts and Heaven

Type of work
+ DVD packaging
++ CD packaging
+++/++++ Album cover

Art Direction/Design
AD/D Carl Rush

Year produced
+/++ 2002
+++/++++ 2001

Specification
+ Use of material: Folded poster
containing a DVD of the movie "Your
Gonna Wake Up One Morning"
Dimension: 28 x 13 mm

+++ ++ Use of material: A jewel case
contained in a metallic antistatic
computer chip bag and stickered
Dimension: 120 x 120 mm
+++/++++ Use of material: Cardboard
CD case
Dimension: 120 mm

Description
+ A commission to produce
packaging for a film "Your Gonna
Wake Up One Morning". It had to
be a cost-effective packaging but at
the same time, represented the film
in some way. A giant "speed (drug)
wrap" was created to contain the DVD
and an invitation to the premier. (The
drug reference is part of the film's plot)
++ The artist (QDF) is space-mad and
wanted his album to represent this
passion. The symbol on the front is a
representation of a continuum space
loop. The silver package gives it the
space vibe.
+++ The image is based on two old
ladies in an edit suite and the chaos
caused by their mixing actions.
++++ The image is a mix of opposites
– pornography and the playfulness of
a dot to dot book.

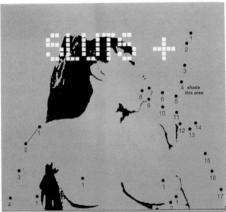

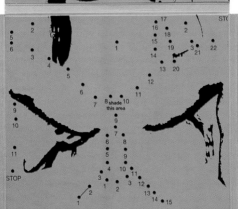

++++

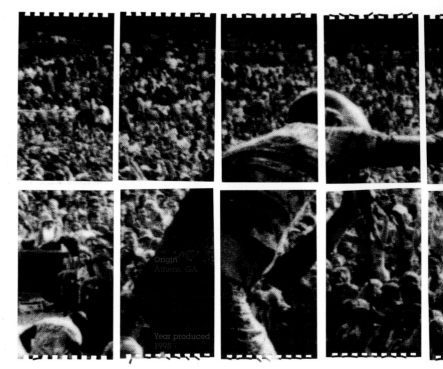

+ Chris Bilheimer

Page
216

Client
R.E.M.

Title of work
R.E.M. UP

Art Direction/Design
Chris Bilheimer

Type of work
CD packaging

Specification
--

Description
This package is a limited-edition cardboard box enclosed in food-grade plastic wrapper. The box contains a 48-page full-colour spiral-bound book. Pages may be torn out and pieced together to form a giant poster.

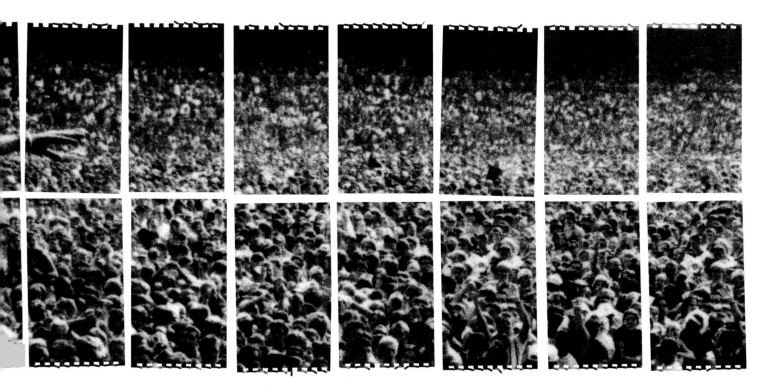

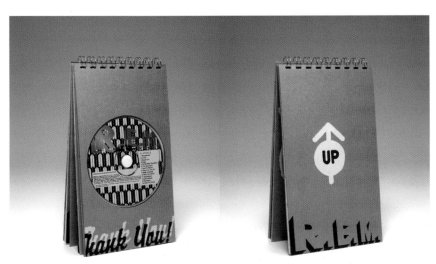

■■■■■■■■■■■I™

+ Chris Bilheimer

Page
218

Client
Man Or Astro-man

Origin
Athens, GA

Title of work
Man Or Astro-man
UFO's And The Men Who Fly Them

Art Direction/Design
Chris Bilheimer

Year produced
1997

Type of work
7" vinyl package

Specification
--

Description
This package 7" sleeve is printed on the reversed side of an uncoated cardboard. The sleeve is die-cut, allowing it to self-close. The sleeve also has die-cut pieces that may be removed and glued into an actual three-dimensional flying saucer. All interior printing is in silver ink.

additional

additional is an open workshop-based project, it mutates according to its contributors and surrounding contexts. additional works within the fields of fashion, graffiti, corporate design, memetics, hacktivism and fields that are yet to be explored.

additional sees itself as parasitic. They do not make their own products, but they adhere to existing things, surfaces and themes. They add onto them and change the purpose for their own aims.

additional declares a red seam to be an independent parasitic textile, while red coloured text spots and red-white security tape become parasitic textures.

Page 128-129

Airside

Airside is a London-based design group started in 1999. They work in every possible medium, and at the moment they are working on knitting, screen-printing, badge-making, pop promos, posters, web sites, brand identities, events, tee-shirts and animations for mobile phones.

They believe that all design is experiential, and so they focus on elegant and efficient interfaces in every medium.

They also believe in the concept of communities, and they create them through their work by inspiring and empowering people. In practice, this means that they like to welcome people and encourage them to make friends with each other.

Page 58-59, 140-141, 144-145

BankerWessel

BankerWessel is a design company based in Stockholm, Sweden. The company was founded in 2001 and is a partnership between Jonas Banker, an illustrator and Ida Wessel, a graphic designer.

Their designs are developed from the varying preconditions of their commissions, and their goal is to enjoy and surprise, not the least ourselves.

Page 172-173

Bark

Bark was established by fellow Royal College of Art graduates, Jason Edwards and Tim Hutchinson in 1995. As a creative partnership, they head up a team of people that have created design briefs for clients as diverse as Rankin, Sony and Nigel Coates.

Bark has been committed to delivering the best-bespoke solutions in line with the needs of the clients. Their understanding of contemporary culture makes them competent in the consultancy and design of branding including packaging, identity and promotion.

"Bark are more than just great designers - they helped us identify some key ideas about our business through the process of creating an identity - like some kind of weird 'graphics psychiatrists...'" Jeff Kirby - Director - Urban Research laboratory Ltd, London

Page 34-37

Bleed

Bleed is a graphic design agency based in Oslo, Norway. It was established in June 2000 by five individuals with background in graphic design, advertising and technology. Today Bleed has twelve employees and stands out both nationally and internationally as a creative design agency that challenges the borders between commercial and art projects. Bleed believes that design must be based on outstanding creative ideas, and that identity, function and values all attribute to make the design work. Based on these values, Bleed applies graphic design as a communication tool and develops innovative design solutions for all channels.

Page 183

Bluemark

Bluemark was established by Toshihiro Saito, Atsuki Kiku-chi, and Akihiro Sato in 2000. Their aim is to provide "high-quality service" in fields such as brand construction, graphic design, web consulting, systems development and publication.

Page 39, 53

BombTheDot

BombTheDot started in 1999 as design-collective of four designers and photographers. The company's main focus is on graphic designs for CD, vinyl and DVD releases as well as concepts and management for prints and packaging (special paper / cardboards, special inks, embossing, cutouts, aluminum and steel-packaging).

Page 56-57

Brandia

In the last twenty years, Brandia has changed some of the most important Portuguese brands. With two hundred and fifty employees and more than fifty areas of expertise, Brandia Novodesign is an Iberian leader and one of the largest brand management companies in Europe. Their philosophy is: The right inspiration and the right partnership for great challenges.

Page 124

BrandIgloo

SecretDiary started almost a year ago and from a simple idea, it developed into an almost ten-month project involving a lot of befriended designers and programmers who helped to realize the project.

During the process of developing and creating the SecretDiary, a company was needed as a base and background for the diary as well as being able to distribute the diary and other products online. Hence BrandIgloo was founded last year as the company behind the SecretDiary. The name BrandIgloo mirrors the basic idea of the SecretDiary, i.e. an igloo is a shelter for closely related people, keeps them warm and protects them from the cold outside. Thus the SecretDiary can be regarded as a "digital shelter" for two people.

Page 212

butterfly-stroke inc.

Since its establishment in 1999, butterfly-stroke inc. has primarily been involved in the creative side of advertising. The company's management and licensing division was established in January 2003 to manage artists and their creations, including character licensing, merchandise planning and production as well as sales.

Aoki Katsunori is the President of butterfly-stroke inc., but he is also involved in the graphic design, art and creative direction of the company. Aoki is a talented creator who has produced successful work within the fields of advertising, graphic art and design as well as filmmaking. His work is known for the use of novel and inspiring visuals, which includes scribble-like drawings, photographs, or computer graphics, all depending on the project. His ability to generate innovative and fresh expressions is reflected in the ever-growing list of clients.

Page 45, 118-119

Chris Bilheimer

Georgia-based designer and photographer is a full-time art director for R.E.M. and he freelances predominantly in the music industry.

Page 216-219

Crush

Established in 1998, Crush is a creative consultancy specializing in design, art direction, concept creations, marketing ideas and advertising solutions for a broad range of clients in the youth, entertainment, music, fashion and publishing industries.

Crush's goal is to blur the boundaries between design and advertising. They aim to focus on original design solutions and are dedicated to diverse creativity and innovation. They design and build marketing concepts and creative campaigns for international brands. They also design for major and independent record labels, and work with publishing houses as designers and consultants. Their client base is growing and they are always looking for new challenges.

Page 213-215

Cuban Council

Cuban Council is a small interactive agency, which specializes in creating superb solutions for web and print. Their main office is located in sunny San Francisco, but they also provide services for clients all over the world. Their areas of expertise range from UI, application and graphic design to iconography, multimedia authoring and complex back-end development. As a group, they take great pride in developing solutions that are fun, usable, technically challenging, visually exciting and / or quirky.

Page 66-67

Curiosity Inc.

Curiosity's president/French designer, Gwenael Nicolass moved to Tokyo thirteen years ago after graduating from RCA in London. He established design company "CURIOSITY" in 1998, and their projects cover various fields from interior, industrial design to package and graphic design. In the cosmetic field, Curiosity designed many perfume bottles worldwide including cosmetic items for Kanebo, Japan.

Page 156-163, 202-203

dossiercreative inc.

dossiercreative inc. is a leading innovative branding design firm, serving a diverse cross section of companies throughout North America. They team with clients to create innovative branding using their proprietary project approach, Creative Intelligence™. This is an essential synergistic process used to generate brands with potent emotional positioning. dossiercreative's brand development services include brand strategy and positioning, brand naming and identity, market research, packaging, launch and promotional materials and three-dimensional branding.

Page 84-85

Doyle Partners

Doyle Partners merges expertise in graphic design, communications, advertising, and marketing in a small studio environment. For almost twenty years, their staff of eight continue to impress a broad roster of clients with design concepts that are active and engaging, implemented with discipline and imagination. This is a studio where identity programs, packaging, magazines, catalogs, books, CDs, installations, signage and environments, and products are created simultaneously for a varied clientele.

The variety of projects keeps the thinking and the work fresh; limiting the number of projects keeps a focus on the needs of the client. What unites them is an honest passion for the power of communication and dialog — and an overriding determination to get the colours right.

Page 204-205

Enric Aguilera

Enric Aguilera is a creative studio from Barcelona founded in 1986. They develope their professional activity in different areas such as graphic design, advertising and packaging.

Page 86-87, 109

Époxy

Founded in 1992, Époxy is a cutting-edge creative shop specializing in the technological, cultural and branding sectors. With offices in Montreal and Paris, Époxy offers a full range of services including: advertising, branding, two or three-dimensional graphic design, film or new media animation and interactive communications (mobile telephony, Internet, CD-ROM and DVD-ROM). The company's creative talent has been acknowledged repeatedly both at home and abroad. Époxy's wide-ranging client portfolio features prestigious names as the Daniel Langlois Foundation, Dharma Resorts, Walt Disney, Eyewire/Getty Images, In-terbrew, Labatt Breweries, Avid/Softimage and Sony C.A.L.

Page 102-105, 200-201

Form®

Form® is an award-winning graphic design consultancy based in London. Established by Paul West and Paula Benson in 1991, their company's background focused on creating campaigns for many high profile bands in the music industry, working as art directors and designers of record sleeves, logos, POS and merchandise. Over the years, their client list has expanded dramatically and in addition to the music work, they now receive regular commissions for identity and branding: book, brochure and DVD packaging design, advertising campaigns, websites and moving image projects.

To date, Form® clients have covered the areas of music, sport, fashion, furniture, architecture, interiors, events, TV, film, video, publishing, marketing and PR. They welcome challenges from all areas where clients require an inspiring and creative approach.

Page 40-41, 131

Freitag

In 1993, Daniel and Markus Freitag decided they wanted bags like the messenger gladiators in New York City. Star-ing out of their living room onto the Zurich transit route, they were inspired and later created individual bags made from recycled truck tarpaulins, seat belts and bicycle inner tubes. The original bag was sewn in that same living room. Today, their bags are designed, cut and assembled at the F-Fabrik, which is still next to the old transit route in Zurich.

Page 178-181

Gabor Palotai Design

Graphic designer Gabor Palotai uses an urban and international idiom to showcase the unique perspective of each assignment, whether it is books, packaging, pictograms, animation, or exhibitions. Palotai has made a highly original contribution to Scandinavian graphic design. He creates powerful and radical aesthetics, a combination of seriousness and playfulness focused on early modernist achievements, especially on the Central European tradition. Palotai's designs are collected in the book "Maximizing the Audience" which comes together with the experimental DVD. He is a member of the Alliance Graphique Internationale (AGI), and he has won a number of prizes for his work, most recently the 2003 Red Dot Design Award. His work has appeared in many publications and some of them are in the permanent collection of the National Museum in Stockholm.

Page 108

groovisions

groovisions is a design group lead by Hiroshi Ito. The name, "groovisions" was given by Yasuharu Konishi of Pizzicato Five in 1991 when Ito produced the art visuals for the Pizzicato Five's World Tour.

Today groovisions is based in Tokyo. They manage a wide range of projects including graphic design, motion graphic, sound installation, fashion design and produced work for global companies in the world. There is an original store in Tokyo where it showcases their products, making them accessible to everyone. groovisions is also known for creating the "Chappie" character.

Page 46-51, 94-97, 114-117

Gunnar Thor Vilhjalmsson

Gunnar Thor Vilhjalmsson is a designer living and working in Reykjavik.

Page 149

Han Seeger

Art Director Hans Seeger's passion for packaging is comparable to someone's passion for fishing or woodworking. She sees packaging design as an opportunity to have fun and to work with like-minded people.

Page 76-77, 176-177

Heads Inc.

Heads Inc. is a design studio based in New York City run by designer So Takahashi. They specialize in art direction, graphic design, package design, product design, as well as several other fields.

Page 78-79

Helena Fruehauf

Helena Fruehauf is a graphic designer based in New York where she is currently working for 2x4, Inc. Born in Germany to Chinese-German parents, she received degrees from the Hochschule der Künste, Bremen and the Hochschule für Gestaltung, Offenbach before relocating to the United States. Upon her graduation from the California Institute of the Arts, she has held positions at Ogilvy and Mather's Brand Integration Group in Los Angeles and New York. She has given lectures at the School of Visual Arts, New York and the University of Missouri, St. Louis among others. Her work has been exhibited and published in Europe as well the United States.

Influenced by her background from diverse ethnicities, one of her main interests lies in fusing Western and non-Western cultures.

Page 126-127

K2design

Giannis Kouroudis graduated from University of Athens with a degree in Graphic Design and Photography. His designs and illustrations have participated in three Biennale exhibitions, as well as being featured in many leading Greek and international journals.

Kouroudis is also the creative director of Korres, responsible for the corporate image. He has won several awards including 1st Packaging Award & Merit for Corporate Identity, Golden Star in Packaging design 1st Award in packaging design and in advertising leaflet 2004 and the EPICA Award for Illustrations. Kouroudis also created graphics for the "Athens 2004 Olympic Games Sports Pictograms".

Page 6-13

karlssonwilker inc.

karlssonwilker inc. is a design studio located in New York City, founded in late 2000 by Icelander Hjalti Karlsson and Jan Wilker from Germany. Before karlssonwilker inc., Karlsson worked with Stefan Sagmeister at Sagmeister Inc., and Wilker ran his own modest design studio in Germany.

Three people now work on all sorts of projects, and their clients include: Warner Bros., Princeton Architectural Press, Universal Music, Phaidon Books, MTV, Adobe, The New York Times Magazine, Capitol Records, L'Oreal, MoMA, MOCA, and few others. Their work has received several awards, and has appeared in various design publications and magazines.

Page 188-189

Kokokumaru

Kokokumaru Co. Ltd. is a Japanese graphic design studio run by Yoshimaru Takahashi. Besides being a graphic designer, Takahashi is also the professor and lecturer of Osaka Seikei University and Osaka University of Art Graduate School respectively. He is a member of JAG-DA, JTA, Tokyo-TDC and New York-TDC, and has won numerous awards at New York Art Directors Club (Silver Award), New York Festival (Bronze Prize), Asia Graphics Awards (Judges Choice) and Japan Typography Association Annual (Best Work). His work includes 'Yoshimaru', 'Medicine's Graffiti', 'Japanese Traditional Medicine's Package 1.2' and the compilation of work and pictures for 'Yoshimaru Takahashi's Design World 1.2.3' by Kosei Fine Arts Publication.

Page 52, 93, 132-133, 196-197

Less Rain

Less Rain is a creative new media agency based in London, Berlin and Vienna, founded in 1997 by Vassilios Alexiou and Lars Eberle. Their services are aimed at producing high quality, concept-driven, tailor-made solutions for an interactive environment, online or offline. Their clients include Red Bull, Scottish Courage, Nike, Mitsubishi, BBC and IMG.

The broader network of Less Rain consists of over thirty members including encompassing designers, programmers, illustrators, sound engineers, photographers, architects and artists.

'Less' is a by-word for the Less Rain ethos: They make designs that are less predictable without being impractical, devising solutions that are desirable and intuitive by changing the way information is navigated. By creating small, perfect worlds, their goal is to deliver new media experiences that are more about intriguing exploration than off-the-shelf effects.

Page 134-135

Lippa Pearce

Lippa Pearce is a multi-disciplinary design consultancy based in London. Founded in 1990 with Domenic Lippa and Harry Pearce as joint Design Directors and Principal Designers, Lippa Pearce has been featured in the D&AD Annual consecutively over the past thirteen years. Their clients include Boots, Halfords, Marks & Spencer, Unilever, TDK Linklaters, Typographic Circle, Aud-ley, Equazen Nutraceuticals and Geronimo Inns. Having worked together for twenty years as designers, Lippa and Pearce have the philosophy that you have to be brave to stand up for what you believe in and ultimately the desire to design.

Page 155

loup.susanne wolf

loup.susanne wolf has established its name firmly during the ten-year history of the enterprise. They are now known for their competent brand development and the management of premium brands.

Additionally, loup.susanne wolf has received acclaim for the development of intelligent packaging concepts for the business-to-business (B2B) branch and specialized trade. Numerous awards bear testimony to the quality of work completed by their designers and packaging engineers in these named market segments, and in the packaging design of high-value gift sets for the business-to-business branch.

Page 88-91

Meem

Meem is the recording and performing name for solo artist/producer/DJ Michael Moebus. All three studio album releases by Meem ('Machines and Records' in 1998, 'Yum Yum and Miffy' in 2000 and 'The Big Hoo-Hah' in 2003) were initially available in limited edition packaging. Due to the nature and scale of these independent releases, these albums were only available locally in Australia. When asked during an interview why he felt the need to present his music in such elaborate packaging, Michael (Meem) replied, "I wasn't interested in creating another jewel case. Most CD art seemed so sterile to me at the time and I just wanted to create something really different and beautiful. I wasn't worried about any negative pre-judgement. I thought the case would establish positive ideas about the music, if anything. The whole product was an output of me as an artist at the time. I'm glad I made the effort to produce something unique and different."Three-di-mensional World (NSW, Australia) – 18th August, 2003

Page 146-147

Milkxhake

Milkxhake is a new Hong Kong based design studio founded by three designers in 2002, specializing in graphic and interactive designs. The founders started from a design project when studying Digital Graphic Communication at Hong Kong Baptist University. Their name 'milkxhake' not only means a glass of drink, it also symbolizes the spirit of 'mixing', which is expressed through their logo.

Page 139

Multistorey

Multistorey is a London-based Design Consultancy, consisting of the core partnership of Rhonda Drakeford and Harry Woodrow. They formed Multistorey in 1997, after meeting on a Graphic Design Degree course at Central St. Martins (where they have both since taught for several years).

They work across many mediums in many different industries. Alongside with their graphic design portfolio, they have also recently launched an interior product range.

Page 92

Musa

Musa is a Portuguese Design Collective. They promote Portuguese designers around the world, and also develop exhibitions and produce products.

Page 138

Non-Format

London-based Non-Format has worked on a range of projects including music packaging for various record labels, illustration for fashion and advertising clients as well as art direction for the monthly music magazine, The Wire.

Page 72-75, 142-143

Nuno Martins

Nuno Martins was born in 1979 in Oporto of Portugal, where he lives and works today. He graduated in Communication Design from Faculdade de Belas Artes do Porto (1998-2003) and studied at Willem de Kooning Academie – Hoogeschool in Rotterdam, Netherlands.

Page 42-43

Ollystudio

Ollystudio is a London-based creative agency that has evolved into a leading design studio since its inception six years ago. Oliver Walker, the founder brings his fifteen years experience with high profile designs, to a team of creatives who work in these disciplines.

Ollystudio has a history of creating adventurous solutions that also communicate with clear messages. In creating close working relationships with clients, Ollystudio ensures a partnership dedicated to reaching a strong concept and a successful result.

For image creation – they take the latest technology to manipulate photography and illustration and create unique images for clients and help define their brands to the visually-aware customers; For art direction – they work with photographers, stylists, artists and other designers to create image campaigns that promote a brand identity; For graphic design – they utilize typography, design skills and design processes and create work in all of the following areas: advertising, books, brochures, web sites, in-store, POS and exhibition design.

Page 54-55, 125, 148

Output

Output is a design studio formed in summer 2002 by three partners. They now employ a versatile design team with a wealth of skill and experience. They combine the vision of a big agency with the efficiency of a compact and re-sponsive team. Their sole aim is to enable their clients to communicate clearly, creatively and effectively.

As well as some of the major names in broadcasting, they also work regularly with clients in all areas of fashion, PR, music, leisure and the arts.

Page 110-111

Pentagram

Founded in 1972 in London, Pentagram is regarded as one of the world's leading multi-disciplinary design consultancies.

With offices in London, New York, San Francisco, Austin (Texas) and Berlin, Pentagram is organized around nineteen practicing designers/ partners/ directors.

Not only do they offer clients a wide range of skills ranging from architecture, interior and exhibition design to graphic and product design, They also provide services to clients worldwide, from small and large companies to multi-nationals, all of whom benefit equally from the full weight of Pentagram's experience and resources.

Page 82-83, 98-101, 106-107

RDYA

Ricardo Drab and Associates was founded in 1996 with energy, intuition and talent. It is a design studio that acts as a communication and art company, combining advertising strategies, design and marketing. Their clients give them the freedom to create and they work to reach their objectives. Their task is to create a particular language for each brand, an essence that runs through the products and messages of a company, which can be immediately identified by the public.

Page 130

Sagmeister Inc.

Born in 1962, Stefan Sagmeister is a native of Bregenz, Austria. He has a MFA in graphic design from the University of Applied Arts in Vienna and a master's degree from Pratt Institute in New York. Following stints at M&Co. in New York and Leo Burnett Hong Kong, where he was Creative Director, Sagmeister formed the New York based Sagmeister Inc. in 1993. He has designed graphics and packaging for Rolling Stones, David Byrne, Lou Reed, Aerosmith and Pat Metheny. His work has been nominated four times for the Grammies and has won most international design awards. He currently lives in New York.

Page 68-69, 154

Sartoria Comunicazione

Sartoria is a tight-knit family of artisans and cultural architects. Sartoria does not make traditional suits. After all, one size does not fit all. Sartoria tailors custom fit images with speed and accuracy. From designing multimedia communication tools to refining brand strategy; from producing publications, installations and events to practicing guerrilla marketing, Sartoria is fully dedicated to a mission: making clients look good everyday, not just on Sundays.

Page 120-121, 136-137, 150-151

Satellite Design

Satellite Design is a San Francisco based studio founded in 2000 by two principals with very different backgrounds, one designer and one copywriter.

Their fundamental difference lies at the heart of their approach to communication solutions.

They are print-oriented, specializing in identity, print collateral and packaging. They work with some of the world's largest and most visible brands and some tiny start-ups that are still defining their markets, products and identity.

Page 182

Sayuri Studio, Inc.

Sayuri Studio, Inc. was founded in New York in 1998. The multi award-winning studio is known for its unique and highly original work including the revolutionary clear candle. They work for many international clients and they aim to develop concepts and designs that enhance their clients' images, as well as building values to their products.

Page 14-21

Segura Inc.

Founded by Carlos Segura in 1991, Segura Inc. is a multi-faceted design and communications firm that specializes in print, collateral and new media communications. Prior to establishing Segura, Carlos had worked as an art director for many advertising giants such as Young & Rubicam, Ketchum, HCM Marsteler and Bayer Bess Vanderwacker. Besides focusing on their best profession i.e. graphic design, printed advertising, corporate identities and other new media; the company would do any types of creative design through whatever medium is allowed, delivering mes-sages with a distinctive sense of style and simplicity.

Page 174-175

strichpunkt

With a staff of seventeen, strichpunkt concentrates on strategic communication consulting and visual communication, particularly in the fields of corporate design, image media and financial market media. The co-founders of the agency, Kirsten Dietz and Jochen Rädeker (members of the TDC in New York and ADC Germany), place emphasis on high-quality typography and sensitive graphics solutions. Basic criteria in all strichpunkt projects is to have freedom for unusual approaches even when the goals are initially defined and to have fun in every creative process.

Page 190-195, 206-209

Stylo Design

Stylo Design is a creative design consultancy that works for public and private clients on a variety of projects. This includes corporate identity, print, internet, e-commerce, moving image and sound design.

Page 22-31

Sweden Graphics

Sweden Graphics has been around since 1998 and claims to not have an official history. They believe that graphic design has finally started to gain its long needed audience after all these years of new media and I.T. bombardment. They also believe the graphic design world, full of wonderful knowledge about typefaces, colour and compositions, has become an integrated part of the younger generation thanks to the Web. Sweden Graphics' innate understanding of this new graphic world can turn graphic design into a means of expression that is very powerful. This is not simply because they are competent speakers, but they are also competent listeners.

Page 152-153

Templin Brink Design

Launched in 1998 by Joel Templin and Gaby Brink, Templin Brink Design has a vision of intimacy: intimacy of process, where clients are more involved in the creation of the work; intimacy of accountability, where clients know the principals of the firm are doing the work; and intimacy of result, where the work creates a deeper connection between the brand and its audience. Together Templin and Brink have developed an organic working process that inspires creative solutions. They specialize in integrating brands across all media, providing clarity, consistency and stickiness at every touch point with the consumer.

Page 80-81

Teresa & David

Based in Stockholm, the design studio Teresa & David was established in 2001 by Teresa Holmberg, who graduated from Beckmans College of Design, and design journalist David Castenfors. They have broadly worked on graphic identity, motion graphics, book design, magazine and illustration. Variety is important to them, not just to stimulate a high level of creativity, but also to prevent getting stuck into only one certain style. Their work phi-losophy is based on passion, fun, patience and dignity.

Page 44

Traffic

Traffic is a small London based design studio special-izing in design for the music industry. Their types of work include creative projects, packaging, corporate identities and marketing, and their clients include EMI, Parlophone, Polydor, Island, Sony, Paramount.

Page 70-71, 170-171

unfolded

unfolded is a Zurich (Switzerland) based art and design office, with projects ranging from print to digital.

Page 184-187, 210-211

viction:design workshop

viction:design workshop was founded by Victor Cheung in 2001 in Hong Kong. Their workshop provides a broad range of design services to maximize the true potential of their clients, building sophisticated and innovative solutions that enable them to develop deeper and more personalized relationships with their customers.

Page 198-199

Volk

Enrico Bonafede is an Italian graphic designer who has been working on identity, print, typography and packaging since 1996. Enrico was born in Rome, Italy in 1974. He graduated from college in 1995 and began a career in print design. He worked as a graphic designer at a design agency in Rome for five years, and in the summer of 2003, he began his freelance career.

Page 122-123

Warm Rain

Don't forget to say hello/Don't forget to say who we are/ Don't forget to say that we do events, installations, branding, marketing campaigns and graphics/Don't forget to say where the name 'warm rain' comes from/and why 'plinth' is such a lovely word/Don't forget to say the smell of a freshly opened book /and don't forget to say 'how do you do?

Page 168

Yoshie Watanabe

Yoshie Watanabe (graphic designer/illustrator) was born in Yamaguchi Prefecture in 1961. She graduated from Yamaguchi University in 1983 and joined Draft Co., Ltd. in 1986. She became involved in D-BROS, a product design project in 1996, and has continued to do so until today. She has received many awards including the JAGDA New Designer Award (1995), Tokyo ADC Award (1997 and 1999) and Gold Prize of NY ADC (2002).

Page 38, 166-167, 169

Zion Graphics

Zion Graphics was founded in 2002 and is specialized in design for the music and fashion industry. Zion Graphics works with clients such as Sony Music, EMI, Peak Performance and J. Lindeberg. Formally, Ricky Tillblad also runs the design companies, F+ and 24HR.

Page 60-65

ACKNOWLEDGEMENTS

We would like to thank all the designers and companies who made a significant contribution to the compilation of this book. Without them, this project would not have been possible. I am also very grateful to many other people whose names do not appear on the credits but who made specific input and continuous support the whole time.

The interviews are greatly benefited from additional documents and conversations, kindly made available by Giannis Kouroudis at K2design, Sayuri Shoji at Sayuri Studio Inc. and Tom Lancaster at Stylo Design.

Rebecca Lai, CC Lau, Jeanie Wong and Agnes Wong offered much help in the editing and interviews. We would like to thank Alex Chan, Hung, Rosena, Stanley Mu, Shaida Lai and all the producers for their invaluable assistance throughout. Its successful completion also owes a great deal to many professionals in the creative industry who have given us precious insights and comments.

Victionary

FUTURE EDITIONS

If you would like to contribute to the next edition of Victionary,
please email us your details to submit@victionary.com